LIGHTING DEVICES

in the National Reference Collection, Parks Canada

E.I. Woodhead, C. Sullivan, and G. Gusset

Studies in Archaeology
Architecture and History

National Historic Parks and Sites Branch
Parks Canada
Environment Canada
1984

©Minister of Supply and Services Canada 1984.

Available in Canada through authorized bookstore agents and other bookstores, or by mail from the Canadian Government Publishing Centre, Supply and Services Canada, Hull, Quebec, Canada K1A 0S9.

En français ce numero s'intitule **Appareils d'éclairage dans la collection de référence d'objets d'archéologie -- Parcs Canada** (n⁰ de catalogue R61-2/9-21F). En vente au Canada par l'entremise de nos agents libraires agréés et autres librairies, ou par la poste au Centre d'édition du gouvernement du Canada, Approvisionnements et Services Canada, Hull, Québec, Canada K1A 0S9.

Price Canada: $5.50
Price other countries: $6.60
Price subject to change without notice.

Catalogue No.: R61-2/9-21E
ISBN: 0-660-11709-6
ISSN: 0821-1027

Published under the authority
of the Minister of the Environment,
Ottawa, 1984.

The last two chapters were translated by Secretary of State.
Editing, layout, and design: Barbara Patterson

The opinions expressed in this report are those of the author and not necessarily those of Environment Canada.

Parks Canada publishes the results of its research in archaeology, architecture and history. A list of titles is available from Research Publications, Parks Canada, 1600 Liverpool Court, Ottawa, Ontario, K1A 1G2.

Cover: Detail of "Lamplight" by Franklin Brownell, 1857-1946. Courtesy of the National Gallery of Canada (gift of the Royal Canadian Academy, 1893).

**Lighting Devices in the National Reference Collection,
Parks Canada**

E.I. Woodhead, C. Sullivan, and G. Gusset

Submitted for publication in 1982 by E.I. Woodhead, C. Sullivan, and G. Gusset, Archaeological Research Division, National Historic Parks and Sites.

INTRODUCTION

This study of lighting devices represents the combined effort of researchers and cataloguers in Parks Canada Archaeology Division, Material Culture Research, each of the authors contributing in the area of his or her particular knowledge. It was conceived primarily to assist archaeologists in the recognition, description, and interpretation of excavated objects through the presentation of artifacts represented in the National Reference Collection in Ottawa. Material in the National Reference Collection is drawn for the most part from Parks Canada archaeological sites, which have been predominantly connected with military occupation in the 18th and 19th centuries. The collection also includes objects known to have been in common use but not excavated from Parks Canada sites, and objects difficult to illustrate using archaeological specimens.

In this study the artifacts have been supplemented where the archaeological examples offer scant or incomplete information. The study does not include the many variations in styles or mechanisms known to have been available but not yet encountered in excavations. For example, gas lighting has been excluded as artifacts related to this mode of lighting were not represented in the collection. Because this work is based on the collection of archaeologically derived examples, it cannot be a comprehensive history of lighting in Canada. Archaeological material generally reflects objects in everyday use and is usually recovered in a fragmentary state. Elaborate and fashionable goods are rarely excavated.

Lighting devices usually form a small part of the artifact assemblage from a site, although a limited representation need not indicate a limited use of artificial lighting on that site. When establishing date ranges for lighting devices the reader should bear in mind that new developments in lighting methods did not necessarily preclude the extended use of previous methods. A lighting device may predate the occupation of a site as many types of lighting seem to have enjoyed prolonged use.

The lighting devices represented in this study range in date from the late-17th century to the mid-20th century. The material has been organized by the principles of operation involved in the various methods of illumination. Generally the lighting devices being recovered archaeologically seem to be those that were inexpensive to purchase and economical to operate. The objects illustrated here are indicative of the methods of artificial lighting commonly used in Canada during the past 300 years.

From earliest times sparks were created by friction. This is the principle of the flint-and-steel method, which was probably the most commonly used technique prior to the 19th century. The sparks resulting from the striking of steel on flint were used to ignite some dry flammable material, known as tinder. The tinder was often some scorched cloth or threads, punk-rotten wood, or wood shavings. The burning tinder was, in turn, used to ignite a fire for heat or a lighting device which would burn more continuously. The tinder, along with the flint and steel, were usually kept together for ready use in a "tinder-box" (Gloag 1955: 476; Russell 1968: Fig. 19).

Flint, for the purpose of creating sparks, is a nominal term. Actually any hard mineral substance could be used, but stones of the quartz family were preferred; chert, agate, or chalcedony were also used. Discarded gun flints would often see subsequent re-usage as a fire flint when chipped or damaged.

Some chemical methods of producing flame were used in the 18th century. Wood slivers were coated at one end with sulphur, which ignited when brought into contact with phosphorus. Further chemical innovations followed in the early-19th century. The first of the modern matches appeared in 1805. These were wooden splints coated at one end with sulphur and tipped with a mixture of potassium chlorate, sugar, and gum arabic, a combination that would burst into flame when dipped into sulphuric acid. The prepared splints and the vial of acid began to replace the familiar tinder-box (Encyclopedia Britannica 1911: Vol. 17).

The first of the practical friction matches appeared in 1827. These wooden matches were about 3 inches (1 inch = 25.4 mm) long and tipped with a formula of antimony sulphide, potassium chlorate, gum, and starch. They were ignited by being drawn through a fold of rough glass paper (Knight 1855: 273).

The prototype of the modern friction match appeared in 1833. These were coated on one end with a formula similar to the above, but tipped with phosphorus. Ordinary or white phosphorus was used originally, but it was so poisonous that it proved hazardous to the health of the workers involved in the manufacture and dangerous to the user. A modified and harmless form of phosphorus that eliminated the serious problems of the earlier formulas was subsequently discovered. These became the standard strike-anywhere matches found in almost every home in North America, be it rich or poor, as the common kitchen match by 1855.

Other formulas were also being offered to the consumer at this time. In the 1840s the wax match was popular. This was a cotton wick, covered with wax and tipped with sulphur and phosphorus. Many of these were sold under the trade name "Vesta" which had been patented in 1832 (Russell 1968: 45).

In 1855 the safety match was introduced, which ignited only when struck on a particular surface. In this version of the friction match the flammable components were divided between the match head and the striking surface as an insurance against accidental ignition.

The friction lighter was developed about 1900. This device consisted of a flint-and-steel arrangement that ignited a cotton wick saturated with a fossil mineral fuel. The "flint" was a soft iron alloyed with cerium (Russell 1968: 40). Sparks were created when this metal was struck with a rough steel surface, such as a file or rasp, which created the required friction. Modern day lighters operate on this principle and still use the iron-cerium alloy for the flints.

Devices for Creating Spark Found in Archaeological Contexts

In 18th century contexts many occurrences have been reported of fire-steels, or "strike-a-lights" as they are frequently called. Even though these are small ferrous metal objects, their survival is due to their having been forged from good quality steel. A variety of forms are to be found in the National Reference Collection. Those based on either an oval loop or a U-shape occur most frequently (Fig. 1).

Two match boxes are represented in the Parks Canada collection; one is a commercial packaging and the other a personal match holder (Figs. 2, 3). There is one example of a friction mineral fuel lighter (Fig. 4). These three examples are unique occurrences from sites in western Canada.

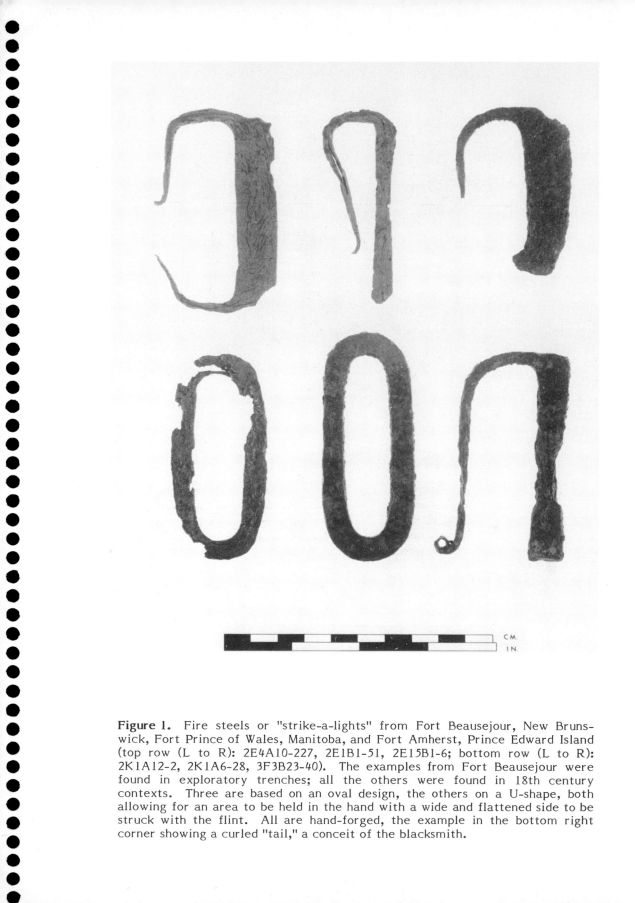

Figure 1. Fire steels or "strike-a-lights" from Fort Beausejour, New Brunswick, Fort Prince of Wales, Manitoba, and Fort Amherst, Prince Edward Island (top row (L to R): 2E4A10-227, 2E1B1-51, 2E15B1-6; bottom row (L to R): 2K1A12-2, 2K1A6-28, 3F3B23-40). The examples from Fort Beausejour were found in exploratory trenches; all the others were found in 18th century contexts. Three are based on an oval design, the others on a U-shape, both allowing for an area to be held in the hand with a wide and flattened side to be struck with the flint. All are hand-forged, the example in the bottom right corner showing a curled "tail," a conceit of the blacksmith.

CM.
IN.

Figure 2. A match box from Fort St. James, British Columbia (3T99A1-2). This artifact was recovered from the general site provenience. The box was made from tinplate and measures 7.3 cm long, 3.9 cm wide, and 2.3 cm high, with a hinged lid. Inside the box against the hinged side is a tray arrangement which would have held one or two matches for easy access. The box was the commercial packaging for a brand of Vespa matches. The raised inscription on the top of the box reads BELL AND BLACK'S WAX VESTA and ...OF LONDON in an oval surrounding the image of a bell. The wax-coated cotton wick Vespa matches were developed in the 1840s, but were replaced in most of this country in the 1850s by the phosphorus-tipped wooden matches.

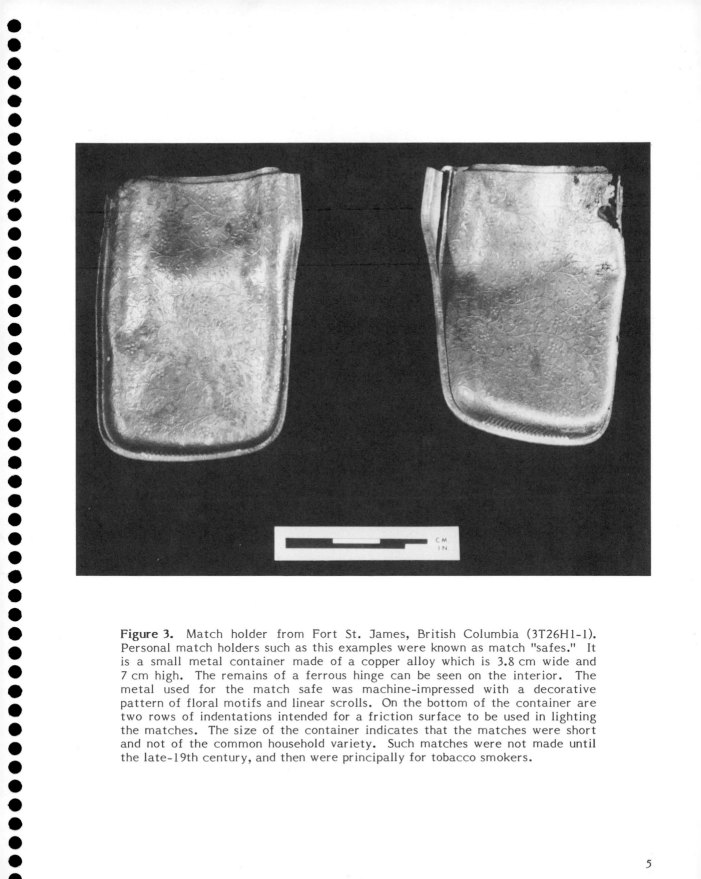

Figure 3. Match holder from Fort St. James, British Columbia (3T26H1-1). Personal match holders such as this examples were known as match "safes." It is a small metal container made of a copper alloy which is 3.8 cm wide and 7 cm high. The remains of a ferrous hinge can be seen on the interior. The metal used for the match safe was machine-impressed with a decorative pattern of floral motifs and linear scrolls. On the bottom of the container are two rows of indentations intended for a friction surface to be used in lighting the matches. The size of the container indicates that the matches were short and not of the common household variety. Such matches were not made until the late-19th century, and then were principally for tobacco smokers.

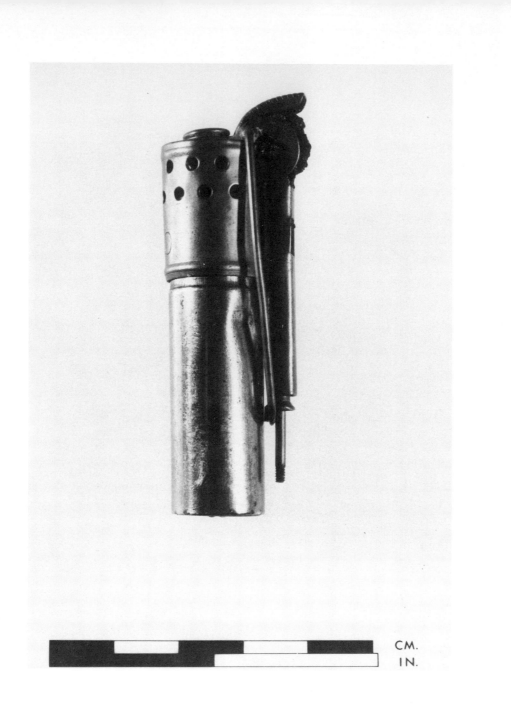

Figure 4. A lighter from Lower Fort Garry, Manitoba (1K99A1-83). The lighter is from the general site provenience. It has been made from brass and bears the marks IMCO and KING 220. These marks have not been identified. The lighter is cylindrical in shape, 6.7 cm long and 1.7 cm in diameter. There is an inner closed cylinder which held the cotton wick and the fuel. The lighter was filled with fuel by the removal of a broad-headed screw in the base. The cotton wick extended out of the top through a small opening. The interior of the fuel reservoir and the cotton wick bear no indication of having been used. A slim rod and wire device on the side of the lighter held the flint in place and exposed it for replacement when necessary.

CANDLES

The candle has been a form of artificial lighting for many centuries, and is still in use today for ceremonial purposes, decorative effects, and emergency lighting. The candle is essentially a self-supporting fuel-enclosed wick. To be self-supporting the fuel must be in a solid form. The fuels used in the making of candles are waxes and solid fatty matter. Traditionally these have been obtained from organic sources: waxes from bees and certain plants such as bayberry, tallow derived from animal fats, and spermaceti from the sperm whale (Hayward 1962: 75). Tallow was the cheapest and most readily available fuel, but it gave a relatively poor light and softened in warm conditions (Russell 1976: 187). Spermaceti gave the brightest light, but used alone it was very brittle; it was also the most expensive alternative (Watkins 1966: 356). In the 19th century palm oil was imported from Africa for candlemaking (Knight 1855: 306-7). Candlemakers often had their own preferred, even secret, formulas which mixed the various ingredients to best suit the purposes and the purses of their clients. The early candlemakers were professionally divided into two groups, the tallow chandlers and the wax chandlers (Hazen 1970: 79). The tallow candles were made by dipping or moulding, whereas the wax candles were made by a pouring method, then rolled to form even cylindrical shapes (Martin 1813: 340-42; Ure 1848: Vol. 1, 252). Candles made in domestic manufacture were predominantly tallow, rendered from the fats of wild or domestic animals. The tin candle mould was a common utensil in early North American households, particularly in remote or rural environments where the animal fats were more easily obtainable than wax or manufactured candles.

In the early-19th century two new solid fuels supplemented the tallows and waxes used for many centuries. The development of scientific chemical analysis led to the isolation of stearine, olein, and margaric acids from organic oils and fats. Stearine was recovered from tallow in 1811 (Kirk and Othner 1967: Vol. IV, 58-59). This fuel gave a brighter and cleaner light than the impure tallow. Paraffin wax, first extracted from crude petroleum in 1850, came into use for candles about 1854.

The wick is an important part of a candle. It must be properly related to the fuel in size and texture to supply neither two much nor too little fuel to the flame. Early candle wicks were simply twisted threads or strands of cotton around which the candle was built. The resultant candles were inefficient and wasteful as they tended to gutter, i.e. to melt on one side allowing the valuable fuel to run away from the flame, or the candle would sputter, the lighted wick extinguishing itself in the melted fuel. The most common problem was the charring of the wick, which choked the flame unless the burnt ends of the wick were removed at frequent intervals with the aid of a candlesnuffer.

Improvements came in the early-19th century with woven forms of wick, which were plaited or braided. These wicks were woven with one thread tighter than the others so that the wick tended to bend as the candle burned, inclining the wick to one side so that it was consumed in the outer part of the flame (Lindsay 1970: 57).

Candles Found in Archaeological Contexts

There is little probability that tallow candles could survive in good condition in normal archaeological contexts. Being of animal extraction they are attractive to small animals, particularly mice (Russell 1976: 188). Both tallow and wax melt at low temperatures and sunlight alone would be sufficient to cause deterioration and loss of their original form so that recognition would be difficult. Nevertheless, there have been several examples retrieved from archaeological excavations. The most interesting examples in the National Reference Collection are those from the underwater excavation of the *Machault,* a French vessel that was sunk in 1760 (Fig. 5).

Figure 5. Candles from the *Machault* (from L to R, top to bottom: 2M8B1-11, 2M8B1-10, 2M8B1-14, 2M8B1-9, 2M16A1-30). Eight tallow candles were recovered in the underwater excavation of the *Machault* which sank in 1760. They had become saponified by their archaeological environment, but were still capable of being lit. The outer saponified layers were whitened, having a chalky appearance, but the melted tallow was of a creamy tan colour. The wicks were twisted strands of unmercerized cotton. These candles were found in close proximity to each other, but no container was evident. They could have been part of a shipment or possibly ship's stores. Candles were usually kept in containers as they were otherwise an invitation to vermin. In 18th century France candles were packaged for sales by the "livre," the number of candles in each parcel depending on their individual size and weight.

Tallow candles made from animal fats were manufactured commercially, but home or domestic manufacture was not uncommon where these products were readily available. In the rural economy of most of 19th century Canada the candle mould was a familiar household utensil. Tallow candles could be made by dipping or in moulds, but the latter was much the less time-consuming and tidier method.

The candle mould was a tubular device made in tinplate with a wide opening at one end and a small opening at the other (Fig. 6). The wick was inserted through the tube and held taut while the melted tallow was poured into the mould. After being allowed to cool and solidify the candle was withdrawn from the mould. The candles could be made one at a time as the tallow was available, but more frequently the mould was made with multiple tubes so that a number of candles could be made in one operation (Lindsay 1970: 41).

Candle Moulds Found in Archaeological Contexts

Most candle moulds were made in tinplate, but as this material has poor survival on archaeological sites there are few represented in artifact collections. Small tubular forms of 2 - 2.5 cm in diameter should be examined carefully as the remains of small tinplate artifacts are usually fragile.

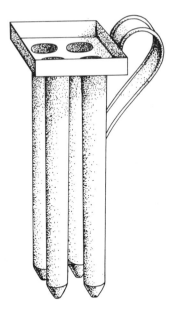

Figure 6. Candle mould. The candle mould was a tubular device with one end narrowed to a conical shape through which the wick was inserted and held taut while the molten fuel was poured into the open end. Most candle moulds were made in tinplate and were generally made in multiples of four, six, eight, or more. A plate across the top held the individual moulds together and also acted as a receptacle for the surplus fuel during the pouring process. Many multiple moulds also had a base plate.

Figure 7. Candle mould from Roma, Prince Edward Island (1F2E2-15). This tinplate candle mould in the National Reference Collection has survived in good condition. Moulds have been found among the artifact remains from other Parks Canada sites but these have collapsed on examination. The context of this mould is dated 1851-1900. The mould is 26 cm long with a diameter 2.3 cm at the open end, tapering to 1.8 cm at the closed end, which narrows from this to a conical point. The longitudinal seam is lapped and soldered. Traces of solder can also be found around both the upper rim and the conical end which suggests that this mould was part of a multiple mould (see Fig. 6).

The purpose of the candlesnuffer was to trim the wicks of candles and early lamps to keep the charred ends of the wick even and clean. This was necessary to keep the flame burning clear and bright, and to keep the loose wick threads from collapsing into the melted fuel and extinguishing the light. The candlesnuffer was a household necessity in the 18th and early-19th century before self-consuming wicks were available. All candlesnuffers were based on a scissor-like operation (Fig. 8). One blade carried a box, the other a vertical flange. To operate this device the charred wick was caught between the blades and cut off, the flange forcing the charred cuttings into the box. The blades were uneven in length, the one with the box being the longer of the two and often having a pointed end. This point was used to straighten the wick threads before cutting them; it was also useful when trying to remove the stub of a candle from its socket (Lindsay 1970: 57-60).

The snuffers were always kept conveniently at hand, and in sight, so that they were often decorated or elegantly fashioned according to the contemporary taste of the period. The snuffers were usually set in or on a holder so that the dirty wick remnants were not scattered (Russell 1968: 33-34).

Candlesnuffers Found in Archaeological Contexts

As candlesnuffers closely resemble scissors in their design and construction, care must be taken not to confuse them. Any example of a scissor-like utensil made in brass will likely be a candlesnuffer, for brass would not hold a cutting edge sharp enough for scissors. The snuffers often were equipped with little peg legs under one of the blades and a finger loop to lift them off the surface when laid down, which were never put on scissors. Other diagnostic differences are to be found in the blades, the box on the longer blade and the flange on the other blade. Any attachment on the blades that may be evidence of these features should be noted.

All the candlesnuffers from Parks Canada sites have been from areas dated 18th or early-19th century. The majority are ferrous metal, but examples of brass candlesnuffers have been found.

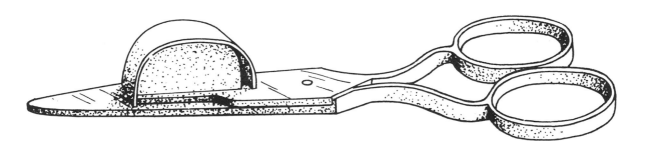

Figure 8. Candlesnuffer. Candlesnuffers worked in a scissor-like motion to trim the charred wicks of candles and lamps. The upper blade bore a vertical plate which forced the charred cuttings into a box mounted on the lower blade.

Figure 9. Candlesnuffer from the *Machault* (2M99A2-11). This candlesnuffer from a 1760 context was cast in brass with the box added to the blade by soldering. The snuffer measures 13.9 cm in length; the width of the blades at the pivot point is 1 cm. The box is 4.5 cm long, 0.9 cm wide, and 2.1 cm high and is semicircular in shape. The flange blade is ornamented on the outer edge.

Figure 10. Handle fragment of brass candlesnuffer from Fort George, Ontario (12H15D2-12). This cast brass handle fragment is from the area of the commandant's quarters dated to the first quarter of the 19th century. The blade measured 1.4 cm wide at the pivot point. The handle shaft is of simple design.

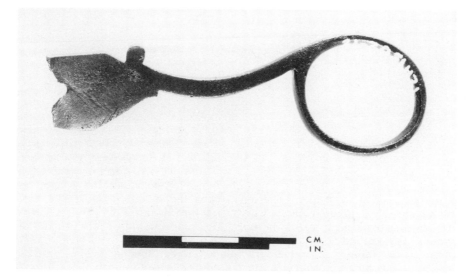

Figure 11. Handle fragment from Roma, Prince Edward Island (1F15M2-11). Made in cast brass, this handle fragment from a candlesnuffer is from a provenience dated between 1747 and 1822. Like the other two examples in brass the shaft and the finger loops are simple in design.

13

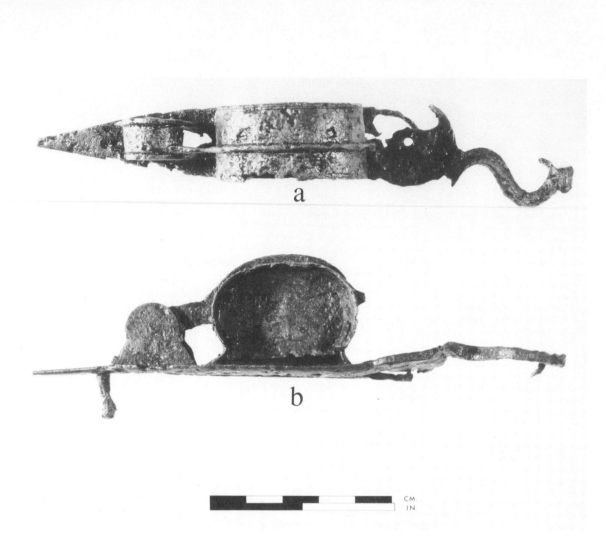

Figure 12. Candlesnuffer from Coteau-du-Lac, Quebec (9G39B1-467). This example was recovered from the ara of the old road east and west of the canal, the strata dated 1779-1815. The ferrous snuffers are 14.7 cm long and 2.5 cm wide at the pivot point. The box is 2.7 cm long, 2.5 cm wide, and 2.9 cm high and is almost round in shape. There is a peg leg mounted under the blade bearing the box. A pivoted blade which fits into a slot cut into the box is mounted on the blade in front of the box. This blade was dropped into the box after the wick ends had been forced in by the flange and it kept these charred fragments from falling out when the blades were opened again. Such a mechanism can be noted on snuffers illustrated in Sheffield catalogues from the early-19th century. The elaborate handles are also typical of this period. (a) Top view. (b) Side view.

CM.
IN.

Figure 13. Candlesnuffer from Coteau-du-Lac, Quebec (9G9K2-23). The snuffer is 14 cm long and 1.2 cm wide at the pivot point. The flange-bearing blade is missing. The box is semicircular and measures 3 cm long, 1.1 cm wide, and 1.9 cm high. This example has been made by forging and welding in ferrous metal. There is a little peg leg under the box.

CM.
IN.

Figure 14. Candlesnuffer from Coteau-du-Lac, Quebec (9G32A1-286). This snuffer is identical with the previous example, but is complete. The measurements are the same. The provenience has been dated between 1800 and 1820, but it is interesting to compare this candlesnuffer, which is made in a ferrous metal, with the brass one found in the excavation of the *Machault* (see Fig. 9).

Figure 15. Candlesnuffer from Coteau-du-Lac, Quebec (9G40H1-9). A candle-snuffer made in ferrous metal, this one exhibits an unusual form of box. The box is rectangular with a domed top. The snuffers are 14 cm long, 1.9 cm wide at the pivot point, and the box is 3.9 cm long, 1.9 cm wide, and 2.6 cm high.

Figure 16. Candlesnuffer from Lower Fort Garry, Manitoba (1K27G13-383). These snuffers are from a latrine, dated to the early-19th century. They are 16 cm long, 2.1 cm wide at the pivot point, with a box measuring 3.6 cm long, 2.1 cm wide, and 2.5 cm high. The box is rectangular with an ornamented top which extends beyond the vertical sides of the box. The handles are decorative. A little peg leg is mounted on the blade which bears the box.

HOLDERS FOR CANDLESNUFFERS

As candlesnuffers were messy things, containing charred and sooty wick cuttings and the drips of fuel which came with them, they were usually kept on a tray or in a holder after use. Such holders could be simple rectangular pans made from sheet metal. More often than not, however, these pans or trays were decorative as they were part of the room furnishings (Fig. 17a). Some of these trays were shaped to fit the snuffers, being wide at one end for the handles and narrow at the other; others were widened at both ends with a narrow mid-section, so that the snuffers could be laid down in either direction. A handle at the centre allowed the snuffer tray to be moved from place to place easily. The trays were made in a variety of metals depending on the taste and the purse of the consumer. Tinplate and brass were commonly used, but sophisticated examples in silver can be seen in museums (Lindsay 1970: 60).

An alternative form of snuffer holder was the upright type (Fig. 17b). These resembled candlesticks except for the socket area. Indeed, examples in cast brass were often made from the same moulds as were used for the base and the shaft of candlesticks as they were made in sets. The socket area of the upright snuffer holder is larger than a candle socket and oblong in cross section. There are openings at the base of the socket for the insertion of the points of the snuffer.

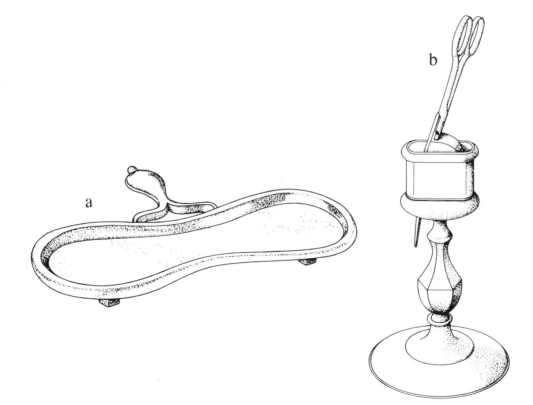

Figure 17. Holders for candlesnuffers. When in use candlesnuffers were contaminated with the messy charred wick ends, so they were placed on trays (a) or in vertical holders (b). The snuffer holders were designed to fit the surrounding decor, i.e. simple and undecorated for utilitarian areas or in the current style for more formal rooms.

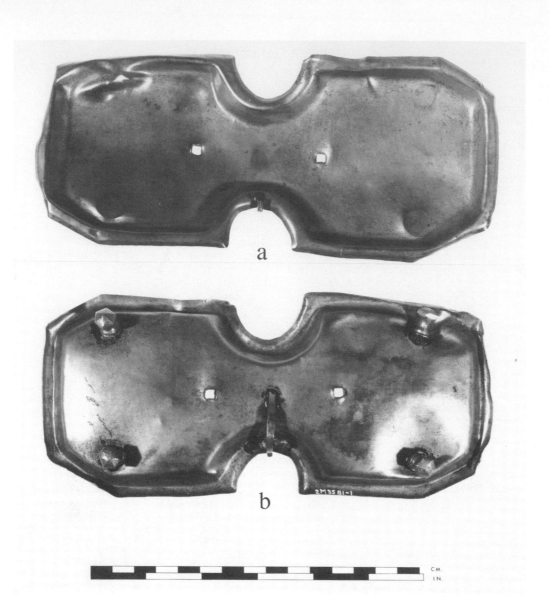

Figure 18. Candlesnuffer tray from the *Machault* dated to 1760 (2M35B1-1). This brass tray type snuffer holder is wider at either end than in the mid-section. The tray is 21 cm long and 9.5 cm wide at the ends. The edge has a raised ridge which follows its contours (a). Two square holes can be seen near the mid-section which held a bracket on which to rest the snuffers. Remains of a cast brass handle can be seen at the mid-section on the underside (b). There were four short legs soldered to the tray which were cast with hexagonal feet. Trays such as this one were common in the 18th century, when brass was a popular metal used for many domestic utensils.

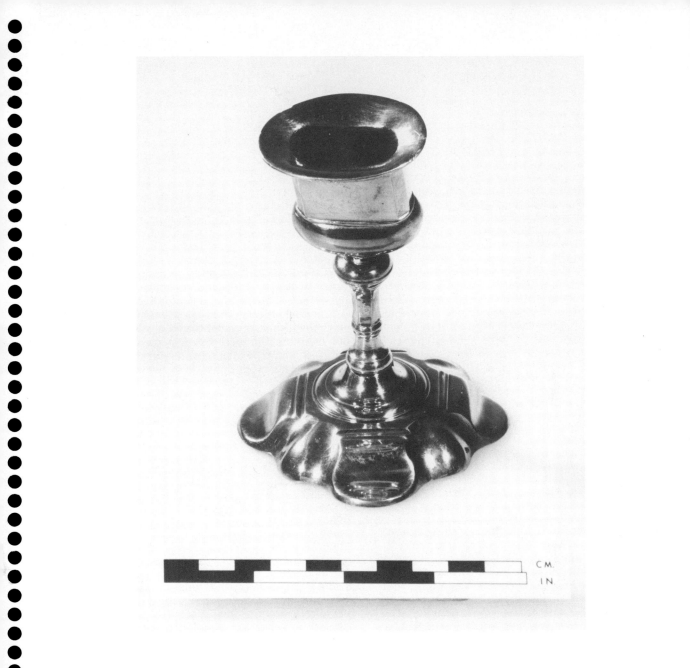

Figure 19. Upright snuffer holder from the *Machault* dated 1760 (2M116C1-4). The holder is 11.5 cm high and the base is 8.6 cm wide. This is an elegant example of an upright snuffer holder made in cast brass. The holder was cast in three pieces, the base, the shaft, and the socket. The base and the shaft are designs which can be found in English pattern books from the 1750s used for candleholders. Although upright snuffer holders are less common than the tray-type, most of those still extant date to the mid-18th century, so that we may suspect they were popular at that time.

CANDLEHOLDERS

Candles could be mounted on almost any horizontal surface, held by a dab of molten wax or by being impaled on a point, such as a protruding nail. However a safer and more decorative solution was to place the candle in a holder designed for this purpose. This could be either a socket arrangement or a point or pricket. The socket is a cylinder into which the base of the candle can be inserted to keep it upright. The socket could be short to hold only the end of the candle or it could be elongated, in the form of a tube, to hold the major part of a candle, with a lifting device to elevate the candle as it burned (Fig. 20b). The latter form was especially functional when tallow candles were used as they tended to soften and bend, or even melt, in warm situations. The candle socket could be mounted on a number of different furnishings to serve the desired lighting conditions. When set on a flat saucer- or tray-like base (Fig. 20a) the style was known as a chamberstick; when mounted on a columnar support with a widened base or foot for balance (Fig. 20c) it was called a candlestick. Other forms using candle sockets were designed to hang on a wall (sconces) or from the ceiling (chandeliers) (Carpenter 1966: 364; Watkins 1966: 357).

Candleholders were sometimes equipped with a handle so that the device could be easily moved or carried about (Russell 1968: 25-26).

As candleholders were part of household furnishings they were subject to a great variety of forms, each conforming to a specific fashion or popular style. It is unnecessary to discuss the innumerable variations here. A study of decorative styles may suggest an appropriate time period, leading to a general-ized dating of the manufacture of a particular candleholder. The candlestick was more sus-ceptible to fashionable taste than the more lowly chamberstick, whose basic form remained virtually unchanged save for varia-tions in the mode of manufacture.

To increase the light intensity from the candle flame, candleholders were often de-signed for more than a single candle. Others had devices, such as reflectors, which inten-sified the light and concentrated it in one direction. Glass or crystal drops were some-times added to the candleholders to increase the effect of the illumination by refraction.

Candleholders were often accompanied by an extinguisher, a cone-shaped device that was put over the flame to exclude the air, putting out the flame without damaging the fragile wick threads or pushing them into the molten fuel.

Candleholders Found in Archaeological Contexts

As candleholders vary a great deal in form they are not easily recognized when in frag-mented condition, as is typically the case for archaeological examples. Metal candleholders appear to occur more frequently than those made in either glass or ceramic materials. Sheet metal chambersticks, the tray-like candleholders, are better represented in the National Reference Collection than candle-sticks. Again, the outstanding examples are from the underwater excavation of the French vessel *Machault,* where exceptional specimens of both types of candleholders were recovered (Figs. 21, 28).

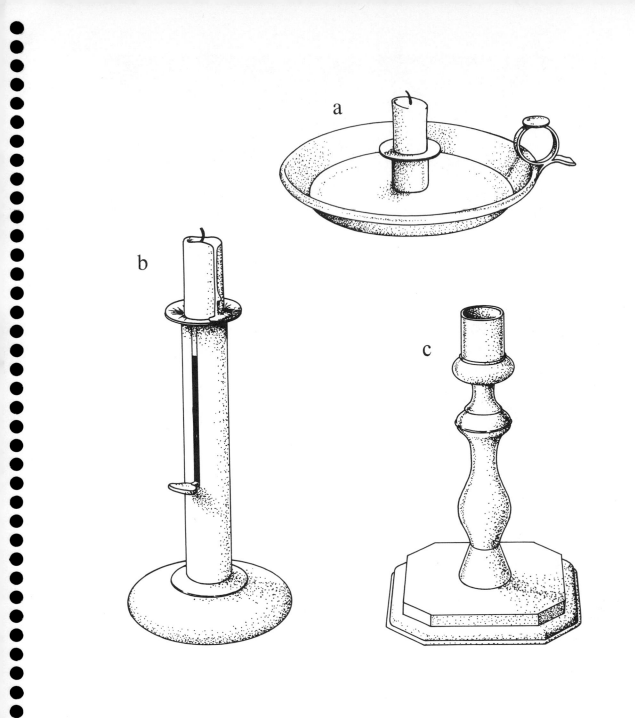

Figure 20. Candleholders. (a) Chamberstick: a candle socket mounted in the centre of a pan or tray, with a side handle mounted on the rim; the handle is generally designed with a thumb rest. (b) Candlestick: a long, slender candle socket with a vertical slit in which is mounted a spring-loaded lifting device which raises the candle; a utilitarian form of candlestick, generally with a simple base. (c) Candlestick: a candle socket mounted on a pedestal; the column and base generally shaped according to the current fashion; a more formal candlestick than (b) and often fashioned in fine metals.

21

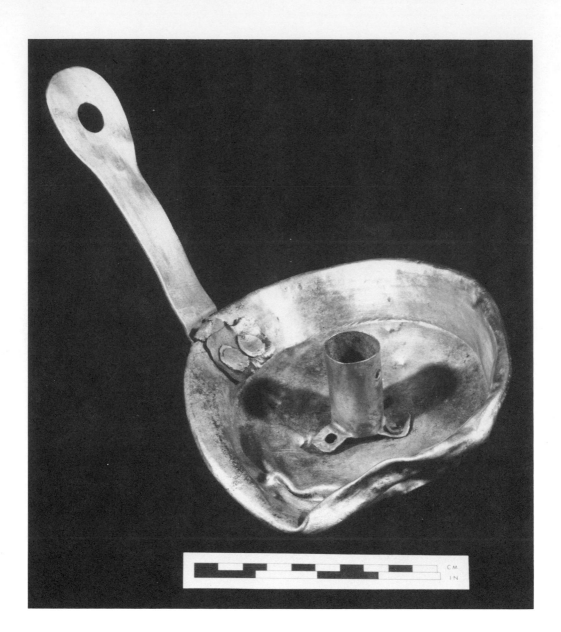

Figure 21. Candleholder recovered from the general provenience of the underwater excavation of the *Machault,* the site dated to 1760 (2M99A2-12). The
candleholder is made in brass, the outer rim diameter of the tray is 15 cm, the
height of the rim is 2.5 cm. The diameter of the candle socket is 2.4 cm. The
rim was reinforced with iron wire for rigidity. The long handle has a large
round hole for suspension of the candleholder when not in use. A mark
consisting of three six-pointed stars appears on the back of the handle below
the hole; it has not been identified. The candleholder was repaired; there are
crude copper rivets holding both the handle and the candle socket in place and a
rough copper plate has been added to the handle repair for further reinforce-
ment, suggesting that the split in the metal occurred prior to its deposit on the
site.

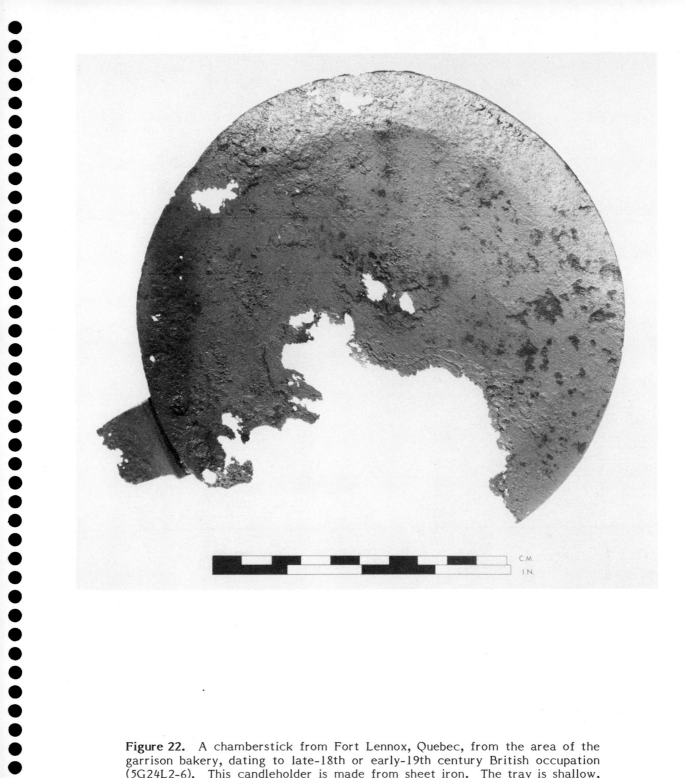

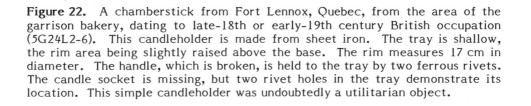

Figure 22. A chamberstick from Fort Lennox, Quebec, from the area of the garrison bakery, dating to late-18th or early-19th century British occupation (5G24L2-6). This candleholder is made from sheet iron. The tray is shallow, the rim area being slightly raised above the base. The rim measures 17 cm in diameter. The handle, which is broken, is held to the tray by two ferrous rivets. The candle socket is missing, but two rivet holes in the tray demonstrate its location. This simple candleholder was undoubtedly a utilitarian object.

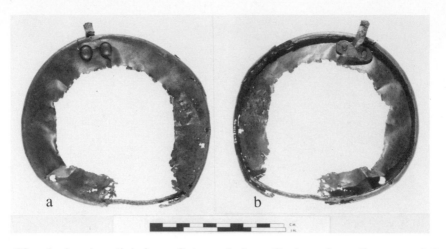

Figure 23. A chamberstick from Coteau-du-Lac, Quebec, from the area of the commandant's quarters, dating to between 1800 and 1820 (9G4C12-45). The candleholder is made in brass, the tray was formed in sheet metal with a wire-rolled rim for rigidity, and the handle, which also has been broken, was cast, leaving a fragment attached to the base by two rivets. This form and style of chamberstick was a common household item, to be found in most late-18th century households. Although its manufacture and assembly are unremarkable, this candleholder has a certain elegance due to the simplicity of the design and the warm tones of the metal. (a) Top view. (b) Bottom view.

Figure 24. A cast brass handle from a chamberstick, from Fort Lennox, Quebec (5G4A8-2). This handle is similar to the one on the previous example of a chamberstick. Made in cast brass it is specifically characteristic of a chamberstick handle in its design. Two rivets and a fragment of sheet brass remain in the area of attachment to the tray. The handle may have broken from the chamberstick before deposit, but there is also the possibility that the chamberstick had suffered damage and the sheet metal of the tray had been salvaged for re-use. Such was often the case in sheet copper or brass objects as these metals were unavailable in North America, except by importation, until domestic manufacture of these metals began in the mid-19th century.

Figure 25. A candle socket, from Fort Lennox, Quebec, from the Navy barracks, dating to the early-19th century (5G22B1-3). This cast brass socket has a ferrous screw-in mechanism to fit it to the candleholder. The interior diameter of the socket is 2.3 cm. This example has a wax deposit on the interior. Screw-in sockets were made for many forms of lighting devices which were built up from castings. The hollow form of the socket was usually cast separately. There was always a market for used brass castings as the metal could be re-melted for new castings, a fact which can help to explain why this material is not found in any great quantity on early North American sites.

Figure 26. Glass candlestick (2L18D23A, 2L18D236). Candleholders in glass material are rarely encountered in forms other than a stick. Archaeologically retrieved specimens can be difficult to identify because of the variety of other stemware forms in glass. There is apparently no evidence of the manufacture of glass candleholders in England before the perfection of the formula for lead glass in the late-17th century (Buckley 1930). Early examples were glassy forms, entirely hollow blown from nozzle to foot, in the Venetian tradition, and very delicate. Later glass candleholders, such as the one illustrated, tended to follow the knobs and swellings familiar on brass and silver candlesticks, resulting in an over-decorated stemware item. Buckley places glass candlesticks between the common household candlesticks of pewter and brass and the expensive silver article. Inventories and period paintings suggest that candlesticks of glass might be used in the public rooms of a home, such as the parlour or dining room, and those of base metals used in kitchens, bedrooms, and the like. This candlestick from the Fortress of Louisbourg is of English lead glass and supports Buckley's statement that in this material, sturdier styles came to be preferred over earlier versions. The flaring lip, which acts as a drip pan at the top of the nozzle, is an unusual feature in a glass candlestick and more closely follows the shape of metal candleholder nozzles (see Fig. 25). This feature also serves to balance visually the heavy and ornate foot. A similar candlestick in the Merseyside County Museums (1979: 65) is dated ca. 1745. (Drawing by D. Kappler.)

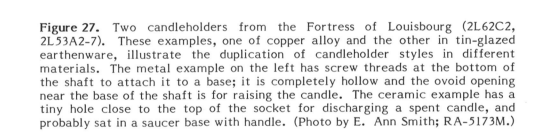

Figure 27. Two candleholders from the Fortress of Louisbourg (2L62C2, 2L53A2-7). These examples, one of copper alloy and the other in tin-glazed earthenware, illustrate the duplication of candleholder styles in different materials. The metal example on the left has screw threads at the bottom of the shaft to attach it to a base; it is completely hollow and the ovoid opening near the base of the shaft is for raising the candle. The ceramic example has a tiny hole close to the top of the socket for discharging a spent candle, and probably sat in a saucer base with handle. (Photo by E. Ann Smith; RA-5173M.)

Figure 28. A pair of cast brass candlesticks from the *Machault,* dated to 1760 (2M99A2-13, 2M116C1-19). The overall height of the candlesticks is 19 cm and the bases are 10.3 cm in width. These fine specimens of 18th century candlesticks are in an extraordinarily fine condition of preservation. The candlesticks were cast in four pieces before assembly: the base, the shaft and the socket in two vertical pieces, and the ring which appears at the junction of the shaft and the base. This ring covers the area which could be rotated, elevating a device in the shaft which would eject the stub of the candle in the socket. This innovation appeared in English candleholders dating to the 1750s. The method of manufacture is also typical of British products of this period, the Continental manufacturers preferring to cast the shafts in one piece. There are no maker's marks on these candlesticks, but they conform stylistically to others known to be from this period in museum collections and those illustrated in period pattern books.

The pan lamp, sometimes referred to as a grease lamp, has been used universally from the earliest times. One of the simplest forms of lighting, it requires only an open shallow vessel, a wick, and some combustible semi-solid or liquid fuel. Known to most primitive societies, these lamps were made from easily obtainable materials. Organic products such as grease, fats, and oils, extracted from wild and domestic animals (deer, bear, beef, or mutton fats, whale and fish oils) and from vegetable sources (olive, sunflower, and palm oils), were the most commonly used fuels. Wicks could be made from pith, moss, plant fibres, or spun threads. In the mid-19th century liquid oleic oil, called lard oil, was isolated from animal fats; it burned with a brighter and longer lasting flame than the unrefined fuels (Lindsay 1970: 50).

The pan lamps produced only a low level of illumination and burned with smoky, smelly flames. The lamps were difficult to light and, once lit, were not efficient; the unconsumed fuel dripped messily from the wick and the charred ends of the wicks needed constant attention to keep a reasonable flame. Often a pick was attached to the pan for cleaning the clogged ends of the threads. The problem of dripping fuel was solved in some models of pan lamps by the addition of a second container below the fuel reservoir to catch the overflowing fluids (Fig. 29). Pan lamps were simple devices and used cheap and easily obtainable fuels. They were messy and gave, at best, a poor light by modern standards but they remained in use until the end of the 19th century in America, being one of the longest lived forms of lighting devices.

When solid or semi-solid fuels such as grease or lard were used, they had to be rendered sufficiently liquid that they could be drawn up the wick. Metal pan lamps were more satisfactory for these fuels than ceramic versions as the heat transferred by the metal from the flame to the fuel kept it fluid. Ceramic pan lamps were more suitable for use with liquid fuels such as vegetable or fish oils.

Many versions of the pan lamp were equipped with hangers and hooks so they could be suspended from the ceiling, wall, or a lamp stand where the light was required. Other variations of the pan lamp included devices that supported the wick in an erect position and methods of covering the fuel pan.

Pan lamps have been used in most parts of the world and were a common form of lighting device in most European countries, especially in rural areas (Watkins 1966: 58). In most instances they were known by a local or national nomenclature. The pan lamps in Canada are often called crusies, reflecting the Scots origin of many early settlers. In Cornwall, England, they were called "chills"; other names for them were slut lamps, "judies," "kays," frog lamps, etc. In some cases the names reflect a particular adaptation of the pan lamp: the Betty lamp refers to a model in which the pan contains a wick holder and the fuel reservoir is covered; the Phoebe lamp has a double base, the lower one to catch the drippings (Russell 1968: 48-54).

Float lamps were similar to pan lamps in that a shallow container acted as a fuel reservoir but, instead of being filled with fuel, it was filled with water and only a small quantity of liquid fuel was floated on the top of the water. A wick was laid on the surface of these liquids, one end being held above the surface by a wood or cork "float." The lamp was self-extinguishing as the flame was doused in the water when the fuel was exhausted.

Pan Lamps Found in Archaeological Contexts

The most typical form of pan lamp was the shallow pan with an extension of the rim to form a wick channel (Fig. 30). Many were suspended from hooks attached to a lug on the rim of the lamp opposite the wick channel. The addition of the second pan, below the fuel pan, was a feature common to many North American pan lamps. Most of the pan lamps found on archaeological sites were made of forged iron, which has been substantial enough to ensure a good rate of survival. In all instances the pan lamps in the National Reference Collection derive from 18th century contexts. One ceramic example has been recovered from a late-17th century underwater site.

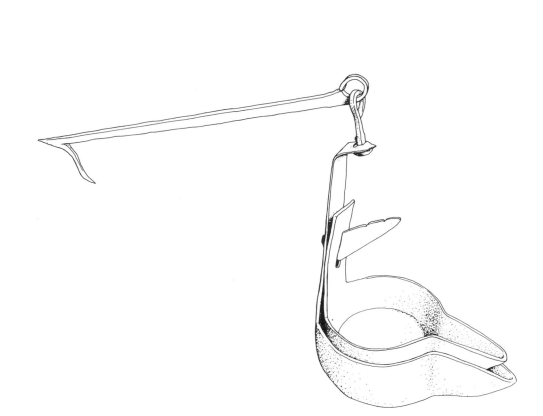

Figure 29. Pan lamp. The illustration depicts the type most frequently encountered, made from forged iron, with a hanging bracket. A drip pan is suspended from the hanger by a swivel device and bears a grooved hook from which the fuel pan is hung. The fuel pan is designed with a broad deep channel for the wick, matched below in the drip pan to catch the unconsumed fuel. The grooved hook allows the fuel pan to be adjusted according to the level of the fuel. (Drawing by D. Kappler.)

30

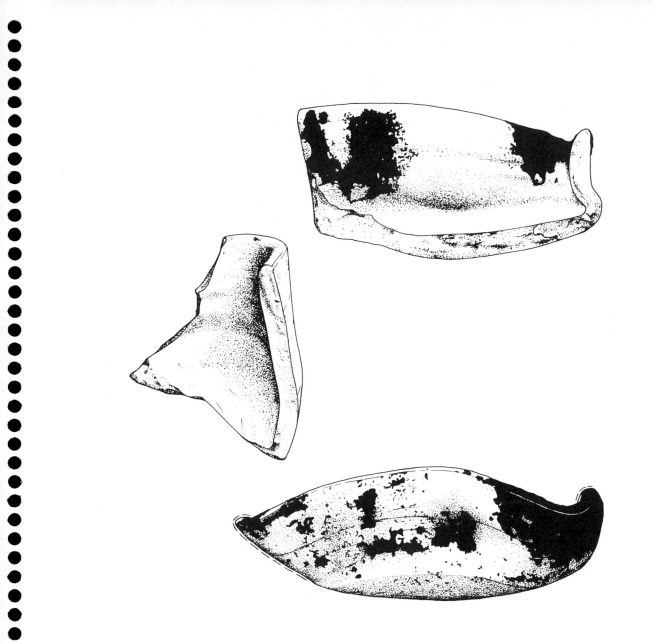

Figure 30. Terra cotta oil lamp fragments (18M38M23-1). This lamp is a type of pan lamp, but is shaped rather like a shallow round-bottomed bowl, with the rim forming several open channels, each one intended to hold a wick. The material is highly micaceous, coarse, red terra cotta, common in the south of France and the Iberian peninsula. This type was identified with the pottery centres of Mérida in southwestern Spain and was made beginning in the 18th century (Hurst 1977: 96; Hurst 1977: pers. comm.). It was made on a wheel, then the rim was folded up and the channels were formed with a finger. Note the traces of burnt oil in several of the channels. The fuel was probably animal or vegetable oil. The lamp was recovered from the wreckage of the *HMS Sapphire,* a British frigate that sank after a battle with the French in 1696, in the port of Bay Bulls near the eastern point of Newfoundland. (Drawings by D. Kappler.)

Figure 31. Pan lamp from Fort Beausejour (2E17P6-59). The overall length of the pan and the wick channel is 11.6 cm. The bowl of the pan and the wick channel are almost equal in size. This lamp is from the officers' quarters, dated 1751-1833, occupied by French and/or British armies.

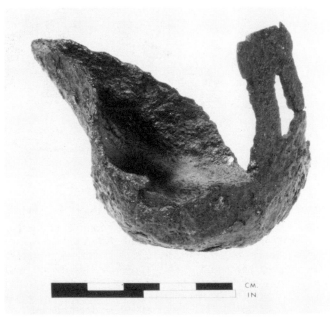

Figure 32. Pan lamp from Fort Beausejour (2E20G20-29). A fuel pan from a pan lamp with overall length of 10.5 cm. The wick channel is sharply upturned in relationship to the bowl of the pan. The lamp fragment was found in an area occupied by the French and/or the British, 1751-68.

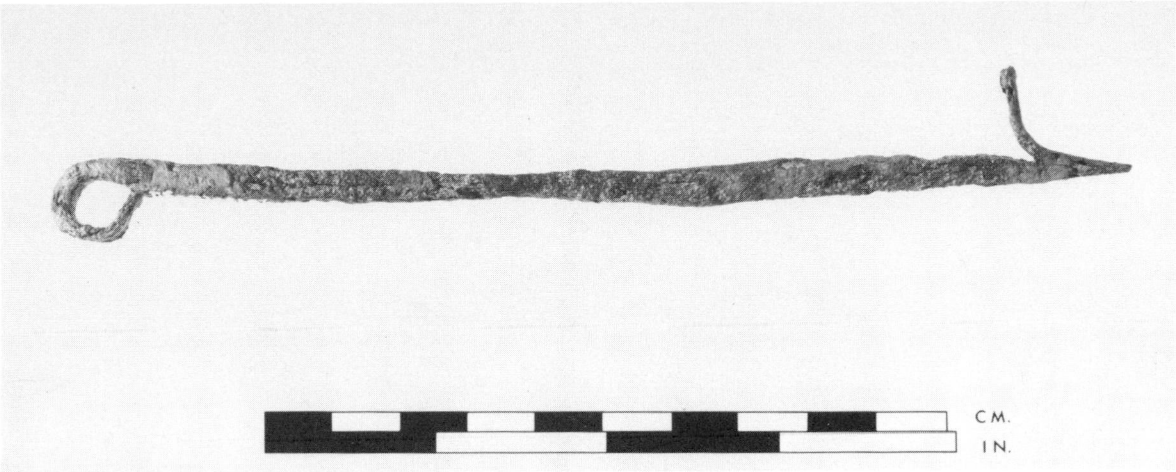

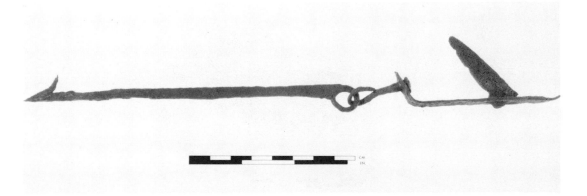

Figure 33. Pan lamp bracket from Fort Beausejour (2E13F10-58). The bracket is made from forged iron rod with an eye at one end for a swivel device and a point at the other for driving into position. The bracket is 16 cm long. It was recovered from a casemate used by both French and English between 1751 and 1768.

Figure 34. Pan lamp bracket from Fort Beausejour (2E13K6-103). Forged iron bracket has a swivel device attached with fragments of a drip pan remaining which bear a grooved hook to accommodate a fuel pan. This bracket is 16 cm long; the swivel is 3 cm in length. It was recovered from an area occupied first by the French (1751-55), then by the English (1755-68).

Figure 35. Pan lamp and bracket from Fort Beausejour (2E20G20-29 and 2E13K6-103). Although not necessarily associated originally, two artifacts are assembled here to demonstrate the relationship of the parts of a typical pan lamp. A major portion of the drip pan has been lost due to corrosion.

SPOUT LAMPS

The spout lamp appears to have been a refinement of the pan lamp in which the wick channel was replaced by a tubular device enclosing the wick, and the fuel reservoir was a closed container (Fig. 36). Nevertheless the spout lamp operated on the same principle as the pan lamp and used virtually the same fuels, lard being preferred in most situations. The wicks were twisted threads or pieces of cloth, which were pulled through the tube, or spout. The form of the spout lamp somewhat resembles a teapot. Frequently a second container was placed below the reservoir, with a curved gutter or trough-like projection under the spout to catch the inevitable drip (Watkins 1966: 358).

Although versions of the classical Mediterranean spout lamp cast in brass or bronze were popular for a time in North America, most spout lamps were made in sheet metal such as brass or iron, and particularly tinplate. Tinware was easily and cheaply made, yet durable and efficient.

There were numerous versions of the spout lamp, some having two or more spouts for a brighter light. These large lamps were used in work areas, meeting houses, and churches. Some historical examples have as many as 10 spouts. Many spout lamps were designed to be suspended, others were built with a pedestal, and there were correspondingly many local or particular names for these (Watkins 1966: 359). Small lamps with no drip-collecting device and fat upright wick tubes, with a hook opposite, were often used as miners' lamps. The Cape Cod or Kyal lamp had a drip-collecting container below the fuel reservoir; the Flemish spout lamp, usually made in brass, had a domed cover, a drip channel below the spout, and stood on a tall pedestal with a weighted base; the Lucerna was the Mediterranean type, in cast brass, with one to four wicks, and was more suitable for use with vegetable oils.

Spout Lamps Found in Archaeological Contexts

Only one spout lamp is in the National Reference Collection, that being a tinplate example from Lower Fort Garry, Manitoba. As tinplate is a material that deteriorates relatively rapidly in buried conditions, and the spout lamp was most often made in this material, there could be many more examples that remain unidentified, or unidentifiable, from the scanty remains. Cylindrical tin containers should be closely examined for evidence that may indicate use as a pan or spout lamp.

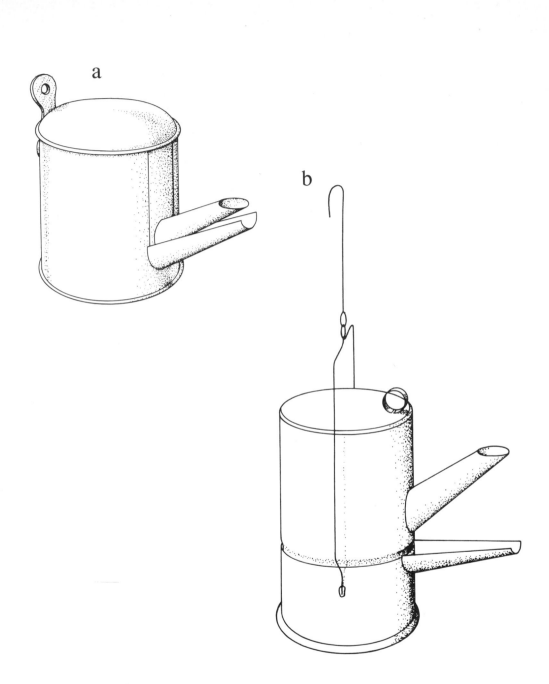

Figure 36. Spout lamp. Many ingenious forms of spout lamps exist; illustrated here are two popular and successful versions. In both examples the lamp has two receptacles; the upper portion was the fuel reservoir with a spout for the wick; the lower portion was the drip pan with a trough or channel mounted below the spout of the upper vessel. The lamps could be set on a surface or suspended; (a) has a lug mounted on the drip pan opposite the channel from which the lamp could be hung; (b) has a bail-type handle for suspension. (Drawings by D. Kappler.)

Figure 37. Tinplate spout lamp from Lower Fort Garry, Manitoba (1K4A1-1648 and 1K5B1-3045). The upper portion of the lamp was recovered from an area of fill, dated ca. 1850, in the vicinity of the canteen. The vessel is 6.4 cm high with an original diameter of 7.3 cm. Before conservation the spout was filled with a fragment of twisted fabric. The contents of the vessel were analysed and proved to be lard. There was no evidence of a handle on the vessel. These features led to its identification as a spout or lard lamp; it might otherwise have been identified as a small beverage container. The lower portion of the lamp was found in the blacksmith's shop, dated 1835-77. It has a diameter of 7.5 cm and is 6 cm in height. The rim of the vessel is wired for strength but cut to allow the removal of a U-shaped portion of the sides. There are traces of solder on the edges of the cut where the drip channel had been attached. Remains of a sheet metal handle or lug are attached to the rim opposite the cut. Distortions of the fragile tinplate do not permit an accurate reassembly. In its original condition the upper portion of the lamp would have set into the lower part, the spout resting above the channel of the drip pan.

VERTICAL WICK LAMPS

Vertical wick lamps have an enclosed fuel reservoir (font) and an opening at the top for a burner. The main difference between spout lamps and vertical wick lamps is that the fuel used in the latter must be in a liquid form to facilitate feeding of the wick by capillarity. Fuels can be made fluid by refining processes that remove the solids, or by being contained in a font that allows heat from the flame to be conducted through metal elements in the lamp to the fuel. Vertical wick lamps evolved from an infrequently used but effective form of artificial illumination to lighting that every-one owned and used.

Modern lighting is considered to have be-gun with Ami Argand's English patent of 1784 (Russell 1968: 75-78). His attempts to im-prove lighting produced a lamp burner with a wick sandwiched between an inner and outer tube, a chimney to enclose the flame, and a metal body with enclosed font and a feeding system that dates to the 16th century. En-closed lamp fonts and vertical wick burners also pre-date the Argand lamp, so these ele-ments were not new with his patent; however the origin of each aspect of the lamp is not known. Following Argand's patent, improve-ments in burners, rearrangement of lamp parts, and development of mechanical devices and additions were made in an attempt to produce lamps that would provide a reliable source of light using a readily available and inexpensive fuel. Early vertical wick lamps were intended to burn whale oil fuel, but the indeterminate supply and fluctuations in price of this fuel limited use to some coastal centres and to the more affluent (Russell 1968: 67). Therefore lamps were designed to permit the burning of different types of fuel depending on availability. A popular form of lamp beginning in the 1840s was the solar lamp, with a metal body that allowed the use of different types of fuels, including lard (Russell 1968: 123-29). In the United States, solar and other types of metal lard lamps continued to be used in isolated areas whereas camphene and gas appear to have been the main means of lighting in urban centres (Rosenberg 1969: 276, 278) until kerosene rendered all other fuels and lamp types obso-lete. However, from the archaeological evi-dence new developments in lighting appear not to have had much impact in Canada, and

illumination by candles and pan lamps seems to have prevailed until the late 1850s at least.

The illuminating properties of coal had been observed for some time before experi-mentation with hydrocarbons for the purpose of lighting was begun. In 1850 James Young was granted an English patent for a process of distilling coal and the products of this distilla-tion, one of which he called paraffine oil (Russell 1968: 134). An American patent for the method and the oil was awarded him in 1852 (Russell 1968: 131-40). Abraham Gesner, in 1846, obtained an oil, which he called kerosene, using a different process of coal distillation; however he did not patent his invention until 1854. Kerosene and paraffine oil were both available in Canada during the 1850s, but were too expensive for general use (Russell 1968: 135). The discoveries that the constituents of petroleum were similar to those of distilled coal and that petroleum could be extracted from the ground by drilling as for salt (Bishop 1967: 463) led to explora-tion for and refining of petroleum. But coal oils took some time to replace other illumina-ting fuels -- the advertisements of merchants and manufacturers of oils and lamps who sub-scribed to the Canada Directory of 1857-58 suggest that even 7 years after the original English patent, in major Canadian and Ameri-can cities the demand for coal oil was minimal (see Lovell 1857: 1242, 1257, 1262, 1424, 1448, 1480), an exception being a New York firm's full page advertisement proclaiming the bril-liant light produced when kerosene is burned in "all the ordinary Solar and Hand lamps" (Lovell 1857: 1479; reproduced in Russell 1968: 135). The proliferation of petroleum wells and re-fineries beginning at about this time lowered the price of kerosene and created confidence in its availability; the inherent benefits of the fuel for lighting encouraged its use. Russell (1968: 131) concluded that by 1864 kerosene was the most commonly used lamp fuel in North America. Archaeologically, kerosene lamp parts on Canadian sites can be viewed as a real time marker beginning in the 1860s, when kerosene was being used by all social strata and in all geographical locations.

Older types of lamps could be adapted to burn the new fuel by replacing the burner and chimney or, in the case of some metal lamps, by modifying the body (Russell 1968: 126, 129).

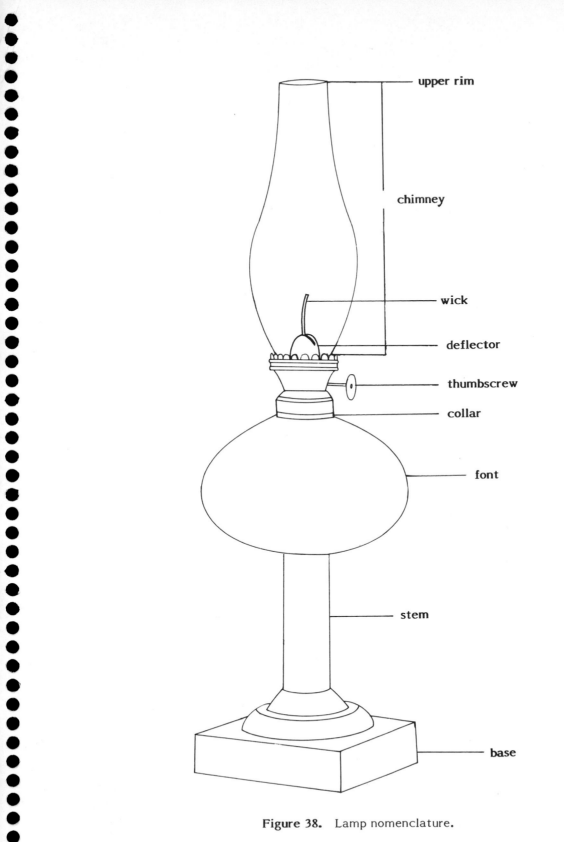

upper rim

chimney

wick

deflector

thumbscrew

collar

font

stem

base

Figure 38. Lamp nomenclature.

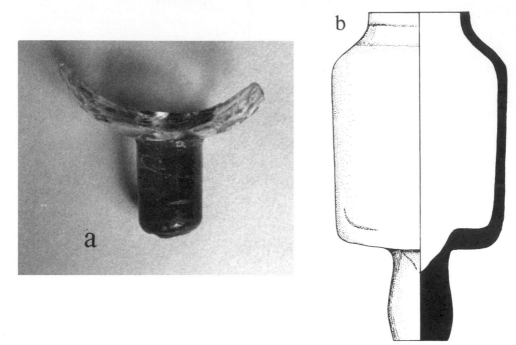

Figure 39. Glass peg lamps (3L11E2 and private collection). The peg lamp consists of a fuel reservoir and a projection or peg, and requires a candleholder or other means to seat and steady it while in use. The origin of the enclosed font is not known; however, non-metallic fuel reservoirs require a fuel with low viscosity at room temperature. The earliest glass lamps were commonly used for burning whale oil, although other liquid fuels were used as well. Stemmed lamps were in use by the end of the 18th century (Russell 1968: 67), and peg lamps are known to date earlier than this. The peg lamp may be the precursor of the stemmed glass lamp which includes a foot. The fragment of peg lamp (a) is from the Fortress of Louisbourg, and dates to about the mid-18th century. It is probably of English manufacture, made of colourless lead glass, and has a pattern-moulded fuel reservoir with a solid peg. There appears to be a pontil mark on the flattish base of the peg; the object has been manufactured in two parts. Were it not for the shape of the font, this item could have passed for a stemmed drinking glass wth plain drawn stem; however, the globular fuel reservoir is not typical of English bowl forms of the 18th century (see Haynes 1959: 194-95). Peg lamps, in English sources referred to as nog or socket lamps, continued to be used in the 19th century and with kerosene. Russell (1968: 141-42) describes one that he has seen as dating no earlier than the 1870s and an English catalogue of 1879 offers them as piano or candlestick lamps embellished with chimney and glass shade (Cuffley 1973: 162). The peg lamp in (b) is a complete example with an opening at the top for a drop burner of the sort used for whale oil. A glass font with a roughened projection, particularly if press moulded and from a late-19th century context, could be a partial lamp, made for insertion into a pedestal base such as is shown in Figure 42 or a base of a different material, and not a true peg lamp (see Cuffley 1973: 188-89). (Photo by Ann Smith; drawing by Susan Laurie-Bourque.)

Russell (1968: 146) feels that because conversion of existing lamps was possible, a distinctive kerosene form took some time to develop. The evidence bears him out if interpreted strictly, but contradicts him in fact.

A British delegation investigating the American system of manufacture, including glass and metal factories in large American cities, published a report on manufacturing in the United States in 1854 (Rosenberg 1969). The delegates found the Boston and Sandwich Glass Factory making large quantities of inexpensive glass lamps, which they call peculiar to that country, decorated by press-moulded patterns, for burning oils and camphene (Rosenberg 1969: 272, 276, 287-89). Lighting devices being made in metal-working factories at the same time were lard lamps, particularly for the western markets (Rosenberg 1969: 276), pedestal stems made of sheet metal to be used with glass and ceramic fonts, and the fixtures used in gas lighting (Rosenberg 1969: 272). Some modern authorities such as Thuro (1976: 81) and Russell (1968: 140) consider the lamp with metal stem and glass font to be an early kerosene form. As James Young's British and American patents for coal oil were only 4 and 2 years old, respectively, at this time, it appears that the technology to produce this lamp was developed before kerosene was widely marketed. Thus it would seem that a lamp shape intended for a different fuel was among the first used for kerosene. Because the lamp and the new fuel became available at about the same time, the two became associated with each other although they were not originally intended to be used together.

Patent information cited by Innes (1976: 312-13) has led him to suggest that the availability of lamps was hastened by the popularity of camphene as a lighting fuel. Press-moulded lamps being made by Boston and Sandwich before 1854 were probably thick-walled camphene and burning fluid lamps which required a fuel reservoir of a particular shape to prevent explosion (see Fig. 40). Kerosene fuel can safely be burned in a font of almost any configuration. American glass houses were manufacturing press-moulded hollow and flat objects by the late 1820s, and by the 1860s could make a wide variety of glassware items by entirely press moulding or by joining together two press-moulded objects made separately.

Two popular glass font shapes for kerosene lamps are the flattened globe and the flattened square, both of which appear on lamps of the late-19th and 20th centuries. Lamps made of ceramic material have not been encountered in the literature or on our sites, but metal lamps were made throughout the 19th century and later (Thuro 1976: 60-67). The Aladdin lamp of the 20th century has an all-metal body. Metal lamps are not a commonly occurring artifact on Canadian archaeological sites. Because the common method of producing lamps involved the use of tinplate, i.e. a thin sheet of tin bonded to ferrous metal, these artifacts do not survive well in the ground. Copper alloys, also commonly used in the manufacture of lamps, have a high scrap metal value so were rarely discarded. Furthermore, lamps made of metal have a longer life expectancy than those of a more fragile material, and so are discarded much less frequently.

Hanging Lamp Brackets

Any number of hooks and other contrivances have appeared throughout history for hanging lighting devices from ceilings and walls. Most of the early light sources were kept within reach so that they could be lit and attended to safely and conveniently, or were made so that they could be easily removed from their position when required. Methods involving pulleys and ropes were used for lowering the large and often heavy fixtures in churches and other large spaces.

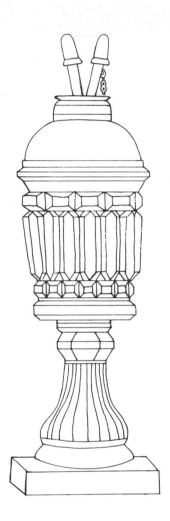

Figure 40. Conjectural drawing of burning fluid lamp. The fonts of burning fluid and camphene lamps, whether of metal or glass material, are high and narrow and taper from top to bottom (Russell 1968: 98). A fuel reservoir with that shape, associated with wick tubes that extend well beyond the height of the burner, was intended to put distance between the flame and the highly explosive fuel (Russell 1968: 102-3). The same shape was used in fonts for glass whale oil lamps, although whale oil fuel reservoirs tend to be thinly blown and less dense than those for camphene and burning fluid. Because whale oil did not require the adoption of a particular prescribed shape, the lamp for burning this fuel can be of almost any shape. As with other forms of glassware, lamps follow the manufacturing and decorative trends of their period of manufacture. Whale oil lamps, made during a time when glassware was commonly made by hand, come in a wide variety of blown shapes. Lamps originally designed for burning whale oil and burning fluid compounds could be converted to kerosene, as evidenced by lamps of this distinctive shape with kerosene burners. Standardization of collar sizes early in the 19th century ensured that burners fit collars interchangeably, so that conversion was simply a matter of changing the burner and adding a chimney. (Drawing by Susan Laurie-Bourque.)

Figure 41. Footed hand lamp (1K1P1-282). This is a very simple variation of what can be quite an elaborate form — technological advances in the American glass industry all through the 19th century permitted the manufacture of glass objects with extravagant ornamentation at a price that even those of modest means could afford. This lamp from Lower Fort Garry has a mould-blown font, to which a press-moulded foot and free-formed handle, now missing, have been applied. The glass is colourless lead metal. Press-moulded bases applied to mouth-blown objects such as salts, candlesticks. and drinking glasses come from a glassmaking tradition that began in England in the late-18th century (Hughes 1968: 314-15), and continued well into the 19th century, even after a complete object could be press moulded in a single operation. This practice allowed for enormously varied objects composed of different pieces, including examples of lamps whose bases are adapted cup plate patterns (McKearin and McKearin 1948: 379; Innes 1976: 269). Thuro (1976: 81-84) would date the illustrated lamp body to early in the kerosene era, based on its turnip- or pear-shaped font, which she has determined to be a departure from earlier glass lamps, although the shape is similar to metal solar lamps of the same period and earlier. Patent information on the thumbwheel of this example establishes the date of manufacture to be 1865-69 (see Fig. 52). Although glass lamps are not often recovered from archaeological sites, this example was found with the glass kerosene lamp of a later period shown in Figure 43. It could have had several years of use before it was broken and has been thrown out with no apparent attempt to extend its life by mending it. Two discarded lamps in the same lot suggest that replacements were easily and cheaply obtained. (Photo by R. Chan; RA-10131B.)

Figure 42. Pedestal base, probably from a lamp (1A23B4-76). Although this opaque white glass pedestal and its variations are very often seen on lamps (see for example, Russell 1968: Fig. 102; Thuro 1976: 89, 109, 111, 123, 124, 135), it could have been the base for a comport or salver. Innes (1976: 248) shows a comport in two sizes and a whale oil lamp with identical bases, dating to 1840-50. Press-moulded bases made in one operation and attached to a top in another were being produced in England in the 1780s (Hughes 1956: 314-15). Although the glass industry had developed the ability to press mould entire objects mechanically in a single process by the late 1820s (McKearin and McKearin 1948: 26), the practice of producing separate parts and joining them continued (Innes 1976: 231-52). This allowed for an almost endless variety of glassware styles by interchanging bases and tops and using different colours of glass. Because the moulded projection is missing from this pedestal, it is impossible to determine the way in which it was attached to its top. An American patent of 1868 allows the manufacture of stems and fonts with screw-threaded glass pegs which were attached to each other by a brass tube with screw impressions. Other patents used different configurations on the glass projections to marry the two parts with brass connectors (see Thuro 1976: 23-24). The joining of opaque pedestals and transparent fonts continued into the late-19th century, although this style seems to have lost its popularity in favour of glass lamps of a single colour, such as is shown in Figure 43. Thuro's examples, cited above, indicate a date of manufacture for the style of base shown here as the late 1860s and 1870s, although an English catalogue of 1885 shows the same item (Cuffley 1973: 79). (Photo by R. Chan; RA-10126B.)

Figure 43. Glass table lamp (1K1P1-281). McKearin and McKearin (1948: 378) define the table lamp size as being about 6 to 10 inches (15.2 to 25.4 cm) in height. Taller versions of the same shape, with a long stem, are called banquet lamps (see Pyne Press 1972: 76-77) and date to the late-19th century (Russell 1968: 256-59). As a form, the table lamp was designed to burn whale oil and burning fluid compounds and was in use in the late-18th century (Russell 1968: 67). Early table lamps tend to maintain the traditional stemware divisions of top, stem, and foot, and some, attributed to American glass houses, very closely resemble stemmed drinking glasses of the same period (see McKearin and McKearin 1948: Pl. 189). Some authorities date the combination of different materials in one table lamp, such as a glass font, metal stem, and marble foot, to the late 1850s, and differently coloured glasses, such as a transparent colourless font joined to an opaque glass pedestal by a brass connector, beginning in the early 1860s (Thuro 1976: 81). Glass table lamps made all of one colour continue from the pre-kerosene period, but the pedestal base replaces the stem-foot divisions. The rectangular shape of the font illustrated bespeaks a date of manufacture well into the time of extensive use of kerosene. It has a press-moulded pedestal stem and a mould-blown font and is made of non-lead glass. A similar table lamp called "Pomona" and dating to 1893 is shown on a trade card in Thuro (1976: 25). This example was found at Lower Fort Garry in the same lot as the footed hand lamp of an earlier period illustrated in Figure 41. (Photo by R. Chan; RA-10725B.)

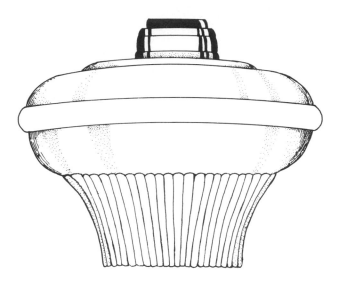

Figure 44. Bracket or hanging lamp font (5G70D10). Bracket lamps are fuel reservoirs intended to be suspended in metal brackets attached to a wall or elsewhere, or hung in a metal frame from a ceiling. The types of holders for bracket lamps, and the possible means of arranging them in holders for one or for multiple lamps is almost endless. In a lamp catalogue of 1859 (Thuro 1976: 18), bracket and hanging lamp fonts are included with two table lamps and a peg lamp whose form is reminiscent of a type of whale oil lamp. Thus, the bracket lamp probably predates the use of kerosene. Both bracket lamps and peg lamps (see Fig. 39) continued to be made throughout the kerosene period and used in similar situations depending on the metal holder. But the flat-bottomed bracket lamp seems to have predominated, possibly because its shape was more versatile and would fit into a variety of holders. Other basal configurations for kerosene bracket lamps include, in addition to the tapering flat-bottomed base shown here, the peg, a solid moulded screw-threaded glass peg which can be screwed into a metal bracket, patented in 1870 (Thuro 1976: 129), and a hole in the base of the lamp which fits onto a metal projection (see Thuro 1976: 74a). The illustrated lamp is a standard early kerosene bracket lamp shape — round-bodied with a flat top and a raised ridge at mid-body to be caught on the rim of a metal cup or basket or contained within metal arms, or a combination of the two. The decorative ribbing on the tapering lower body is also typical of bracket lamps of the period. The lamp was excavated in the navy barracks cookhouse at Ile-aux-Noix, erected in 1816 and levelled in 1870 (Korvemaker 1972: 111). Examples depicted in Thuro (1976: 18, 20, 73-75) indicate a date of manufacture for this shape from the 1850s or earlier to ca. 1870. (Drawing by Susan Laurie-Bourque.)

a

b

Figure 45. Ceiling bracket for hanging lamp from Lower Fort Garry, Manitoba (1K27E3-136). This portion of a lamp bracket, recovered from a latrine, was part of a device for hanging a kerosene lamp. It is made in cast iron and measures 14 cm in diameter. The bracket hung on a hook from the ceiling and the arrangement of three pulleys allowed the lamp to be raised and lowered (see (b)). The lamp would have been lowered for lighting and then raised while lit. It could be lowered again when the lamp was extinguished. The decoration is geometric and highly stylized, a form of decoration popular in the last two decades of the 19th century. Lamp brackets such as this one were used in hallways or in rooms which had high ceilings such as were often found in houses of this period. Illustrations of adjustable hanging lamps similar to this example can be found in many catalogues from the turn of the century. (a) Side view. (b) Top view.

Liquid fuel lamps were similar in principle to spout lamps in that both had fuel reservoirs with wick tubes, the prime difference being in the fuels used. Liquid fuels were able to rise up the wick by capillary action for ignition. The wick therefore was mounted at the top of the lamp. Liquid fuel lamps are characterized by a closed fuel container with a tightly fitting burner containing a vertical wick tube. Liquid fuels included whale oil, some vegetable oils, and refined lard oil, all of which were rendered obsolete by kerosene (coal oil) after its discovery in 1846. Kerosene was in commercial production by 1855 and in widespread use for lighting purposes by the 1860s (Russell 1968: 195). This fuel was safe, efficient, and, most importantly, economical. Many liquid fuel lamps originally designed for use with whale oil, etc., were subsequently converted to kerosene, many by simply replacing the existing burner unit with one designed for kerosene.

Burners for the early liquid oil lamps were simple. A wick tube, usually round, was inserted into a plug device, some as simple as a cork, others made more substantially of brass, which either fitted the neck of the container used as a fuel reservoir or could be securely screwed into a metal collar rigidly fixed to the reservoir. There were many versions of liquid fuel burners, some containing more than one wick to produce a brighter flame. The most successful and widely used fuels until the mid-19th century were lard oil and whale oil. The Argand lamp of the late-18th century was probably the finest lard oil lamp available; this was also the first lamp to take advantage of the benefits of the addition of a lamp chimney.

When kerosene was introduced in the 1850s the burners initially used for the lamps were imported from Europe, but American manufacturers soon started production of their own, based initially on the European design. The first American patent for a burner was issued in 1858 and hundreds of others followed in the next two decades, each with some improvement or change to one or other parts of the burner.

Most burners were made of copper alloy, particularly brass, in sheet metal, wire, and small castings. The sheet metal parts were produced by die-stamping, which cut out, shaped, and perforated the metal as desired. The assembly of the various parts of the burner was by mechanical means, such as rivets, or by soldering.

The burner consisted of a wick tube, held in place in a unit that screwed into a collar on the fuel reservoir, or font (Figs. 38, 46). The wick was raised and lowered in the tube by a mechanical device of toothed wick wheels. These wheels were operated by a thumb wheel on a shaft extending to the exterior of the burner. The wick used on most North American kerosene lamps was a flat woven cotton or asbestos strip (Cuffley 1973: 35). Therefore most wick tubes were rectangular in shape. The majority of kerosene lamps were designed for use with a chimney which both protected the flame and, by the ingenious design of the burner, actually assisted in providing an air current to the wick to produce a brighter light. The burner for chimneyed lamps included a seating for the base of the chimney, the most common forms being either a gallery or an arrangement of prongs (Thuro 1976: 45). Inside the chimney seating were perforations which allowed for the air flow. In many burners a deflector was added to further deflect the air current to the flame. The deflector had a slit, or blaze hole, through which the flame appeared. Kerosene burners which did not have chimneys were likely used inside lanterns, foot warmers, or other devices in which the lamp was used primarily for heat rather than light.

The efficiency of kerosene lamps was increased by adding devices to increase the fuel consumption. One of these was a vapour vent (Russell 1968: 158). This tube, which paralleled the wick tube, released any build-up of fuel vapour within the sealed font. This was not only a safety feature, preventing a potentially hazardous condition, but actually brought additional fuel to the flame.

The burner was securely mounted on the lamp, screwing into a collar that was rigidly fixed to the font. The sizes of the collars for liquid fuel lamps were standardized early in the 19th century, ca. 1825-30, making conversion from early liquid oil lamps to kerosene simple, and allowing burners to be interchangeable and replaceable.

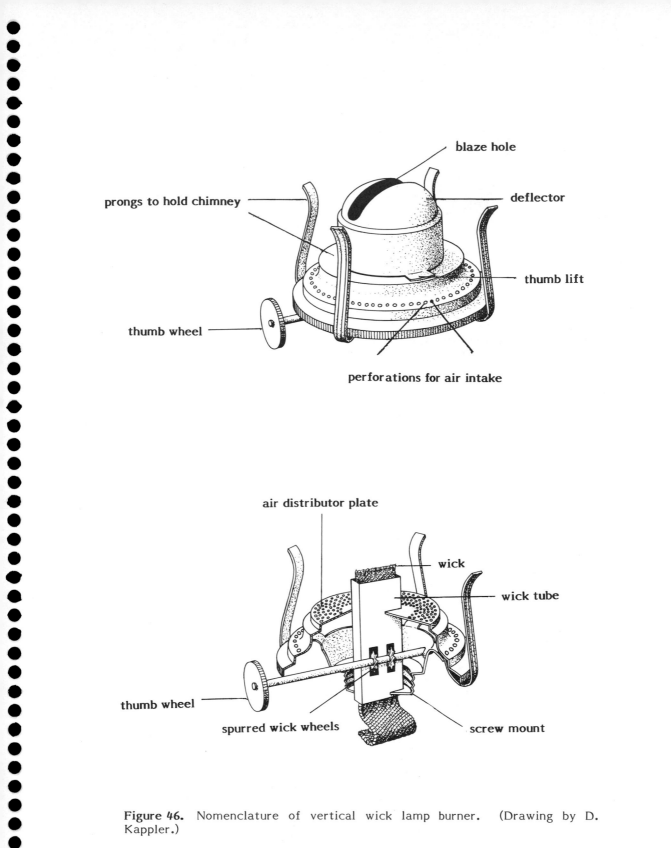

blaze hole

deflector

prongs to hold chimney

thumb lift

thumb wheel

perforations for air intake

air distributor plate

wick

wick tube

thumb wheel

spurred wick wheels

screw mount

Figure 46. Nomenclature of vertical wick lamp burner. (Drawing by D. Kappler.)

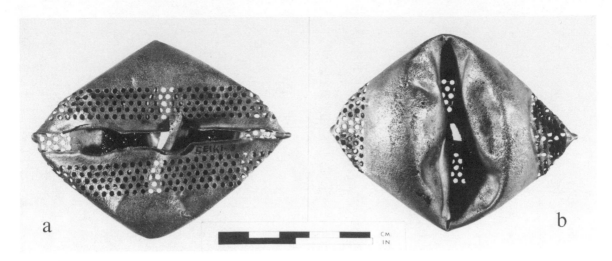

Figure 47. A deflector from a kerosene burner from St. Andrew's Blockhouse, New Brunswick (5E1K1-3). This burner was found in the trenches dug to find the limits of the gun platforms. This deflector is highly domed, with ventilation holes for the air draft around the base. The deflector is 5.6 cm in diameter. Such large deflectors date to the late-19th century, ca. 1890s. (a) Bottom view. (b) Top view.

Kerosene burners are a good example of the American system of manufacture, with the concepts of standardization and interchangeability of parts. Not only were the screw mounts made in standard sizes, but so were the wicks and the chimney seatings. Wicks were made in five sizes: size 0 was 1/2 inch; size 1 or A was 5/8 inch; size B was 7/8 inch; No. 2 was 1 inch; No. 3 or D was 1 1/2 inch. The widths of the wick tubes were made in corresponding dimensions. The sizes of burners and chimneys were as follows: sizes 0 and 1 were 2 1/2 inches in diameter; sizes 2 and 3 were 3 inches in diameter (Wood, Vallance Limited 1911: 102; George Worthington Co. 1916: 413). The diameters of the screw-in collar mounts were 7/7 inches (No. 1), 1 1/4 inches (No. 2), and 1 3/4 inches (No. 3) (Thuro 1976: 39-41).

Many kerosene lamp burners bear makers' marks and patent dates. The makers can often be identified in reference sources, and some company histories can help to date the manufacture of the burners. Patent dates are confusing as they most often refer to only one feature of the burner and not the complete burner itself. This acknowledgement of a patent does not occur on burners took advantage of earlier patents where the patent protection had expired. It is necessary to recognize the variations in the burner and their combinations to differentiate between the parts referred to by the patent information on the burner and those which have become public property. Some significant patent dates are 1860 and 1867 for the vapour vent (Russell 1968: 158); 1868 for the prong-type chimney holder (Thuro 1976: 42); 1861 and 1867 for the hinged burner (Thuro 1976: 38); and 1873 for the prototype of the combination of the hinged deflector and chimney prongs (Russell 1968: 190-91). This last patent had many subsequent variations; the original was named the "Fireside Burner" by its inventors; its subsequent imitators in the 1870s were called the "Eureka," the "Gem," the "Star," the "Eagle," the "Venus," the "Queen Anne," the "Sun," etc. (Cuffley 1973: 46).

Liquid Fuel Lamp Burners Found in Archaeological Contexts

Lamp burners are not abundant on Canadian archaeological sites, but when found the condition of the metal is usually good. Not only does brass survive relatively well, but the fuel oils with which the lamp burners were used are also good metal preservatives. Many burners have survived undistorted and are in perfect working condition even without conservation treatment. Although dating of the burners is of limited use to archaeologists, the recognition of particular styles of burners and the lamps to which they were affixed could be of assistance when reconstructing and refurbishing a historic site.

Figure 48. Lamp burner from Roma, Prince Edward Island (1F2E8-13). This burner is from a provenience associated with the Macdonald store, dated to 1823-1900. The burner bears with it a fragment of the glass font. The burner has a flat wick tube, measuring 1.6 cm wide (0.625 inches). The collar on the font can be dated prior to 1876 by its manufacture. The screw mount is 1.6 cm (0.625 inches) in diameter. There are two patent dates on the thumb wheel, SUNLIGHT PAT. DEC. 14. 1869 and PAT. SEP. 1862. The burner does not appear to have any provision for a lamp chimney, and there is no vapour vent. This lamp burner may have been intended more for providing an open flame for heating rather than for lighting, such as in a medical capacity to provide healing vapours or in a nursery. It could have been used in a lantern, although a glass font was seldom used in this capacity.

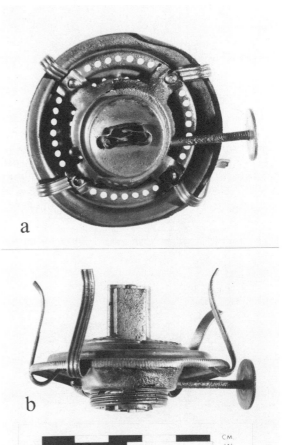

Figure 49. A lamp burner from Fort George, Ontario (12H5B2-3), from a context pre-dating 1882. This burner is the standard type found after 1873 based on a patent which combined the hinged deflector and the prong-type chimney holder (Russell 1968: 190-91). The wick tube is 1.2 cm (0.5 inches) wide, the diameter of the screw collar mount is 2.2 cm (0.875 inches), and the diameter of the chimney seating is 5.7 cm (2.25 inches). There are no marks or patent dates on the burner. This type of burner was common from the 1870s in North America and is still the model used for most burners to the present day. (a) Top view. (b) Side view.

Figure 50. A burner with fragments of a glass chimney attached from Fort George, Ontario (12H18A5-2). This burner was found in an ash pit which was covered by leveling procedures in 1910. Other artifacts found in this pro-

venience have been dated to manufacture pre-1860. The burner is made of cast brass. The actual burner device or the wick holder, which would have fit into the circular opening in the centre, is missing. The chimney holder is a screw-on ring that fits over a flange on the base of the chimney. The outer ring has perforations near the circumference. These features are sufficient to identify this lamp as a Solar lamp. These lamps were manufactured in America from 1839 to 1890; they used olive oil, whale oil, or lard oil as fuels; the fonts were usually all-metal with a steel tube descending from the wick holder into the font to conduct heat to the fuel (Russell 1968: 129).

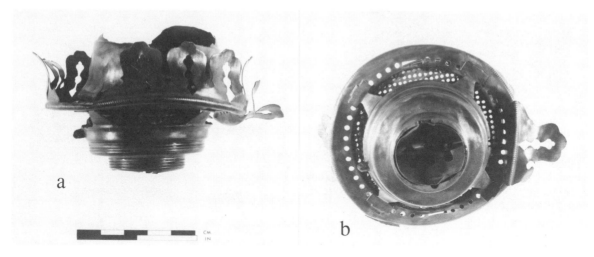

Figure 51. A lamp burner from Lower Fort Garry, Manitoba (1K1L1-64). This burner was recovered from the excavations on the west side of the Big House in a provenience dated to post-1885. It is more ornate and complex than most of the other burners seen from archaeological contexts. The wick tube measures 3.8 cm (1.5 inches) in width on the top of the burner. Underneath, the wick tube is circular and conical. There are three wick wheels to accommodate such a wide wick. There are two sizes of screw mounts so that the burner could be fitted to either size collar, i.e. 3.1 cm and 4.7 cm. The chimney seating has a high gallery with pierced decoration and measures 7.6 cm (3 inches) in diameter. The deflector was highly domed. Large burners of this type were made in the 1890s and the style continued into the first decade of the 20th century (Russell 1968: 238). (a) Side view. (b) Top view.

Figure 52. A burner from Lower Fort Garry, Manitoba (1K1P1-282). This burner, attached to a glass font, was found in the Big House privy. The wick tube measures 1.6 cm in width; the collar mount is 2.2 cm in diameter. The collar on the font is a style that dates prior to 1867. The chimney seating is the gallery type and measures 4.4 cm (1.75 inches) in diameter. There is a hinge on the remaining portion of the gallery allowing the upper part of the burner to be tilted so that the wick could be trimmed without removing the chimney. The first hinged burners were introduced in 1861 (Thuro 1976: 39 and 42). The thumb wheel is marked with both a manufacturer name and a patent date. The manufacturer is identified as Holmes, Booth, and Haydens, a firm which changed its name in 1869. The patent date recorded on the wick wheel is AUG.1.65. With these two pieces of information the manufacture of this burner can be placed between 1865 and 1869.

All the burners in the National Reference Collection were identifiable from standard reference sources. This substantiates the premise of Russell and Thurso that there were certain very popular forms of lamps in Canada.

The various features on the burners all conform to the standardization of parts quoted in these references. Kerosene lamp burners with the smaller sized collar mounts occurred in contexts dated to the 1860s, whereas the larger sized burners could be dated to the last decades of the 19th century and the early-20th century. All the burners from Parks Canada sites identified to date were made in North America; one was manufactured by a Canadian firm but all the others were from the United States. So far no burners found in Canada have been identified as having their origin in Europe or Great Britain.

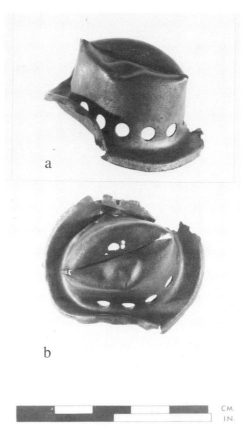

Figure 53. Lamp burner parts from Lower Fort Garry, Manitoba (1K4B1-2005). These burner parts were recovered from the area of the canteen, which had been filled at a later date than the original building. The wick tube is 1.6 cm wide; the diameter of the collar is 2.2 cm. On the back of the ventilator plate is marked PATENTED ... 1868. The burner parts appear to resemble the Collins burner, which was patented in 1865 with improvements in 1868. This model of kerosene burner was in popular use throughout the 1870s (Russell 1968: 163-64; Thuro 1976: 44). It may be noted that this burner was from the same provenience as the spout lamp mentioned earlier in this work.

Figure 54. Deflector from a kerosene burner from Lower Fort Garry, Manitoba (1K11E1-121). The diameter of the base rim of this deflector is 2.5 cm. This small deflector is highly domed with a row of ventilation holes around the base. As the rim is damaged it is not known whether the deflector was of the hinged type or not. (a) Side view. (b) Top view.

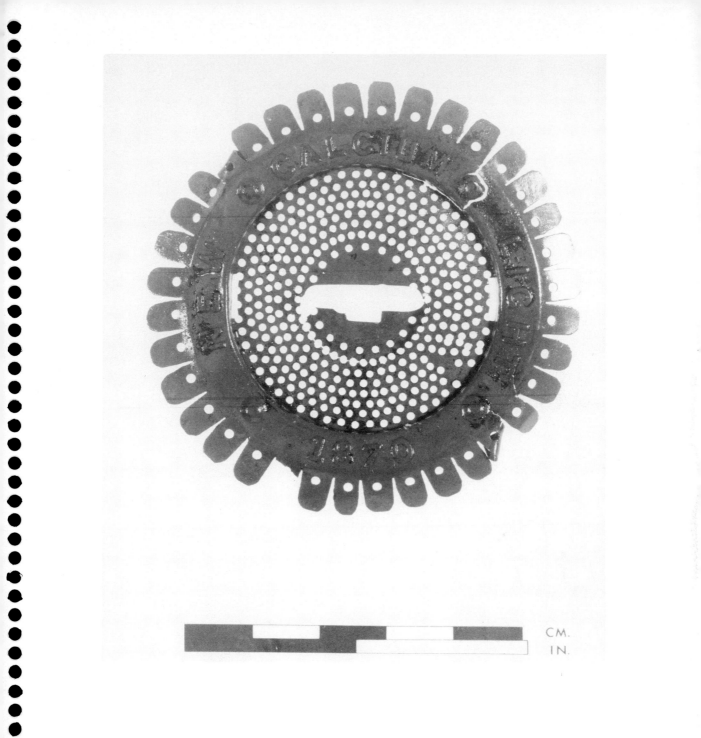

Figure 55. Another kerosene burner part from Lower Fort Garry, Manitoba (1K27G13-381). This ventilator disk is from a latrine. The wick tube is estimated to have been approximately 1.6 cm wide, from the size of the opening on the disk. There is also allowance for a vapour vent. The diagnostic features of this piece are the scalloped circumference and the raised inscription which identifies it as the NEW CALCIUM LIGHT, a burner first manufactured in 1870 (Russell 1968: 189-90).

Figure 56. Kerosene burner from Lower Fort Garry, Manitoba (1K80B1-30 and 1K80B1-75). These two pieces are indeed one artifact. They were recovered from a garbage area between the bastion and the bakehouse which dates to the late-19th or early-20th century. The wick tube width is 2.2 cm; the collar diameter if 3 cm. There is a vapour vent attached to the wick tube. This was a prong-type burner, but the prongs are now missing, although novel slot-like devices for the prongs can be observed. The openings on the ventilation plate reveal that there was a hinged deflector, which is now missing. The thumb wheel drives three wick wheels. There are two patent dates on the thumb wheel, PAT.JAN.16.83 and FEB.13.73. The style of burner is based on Atwood's 1873 "Fireside" burner which was the prototype of many successors (Russell 1968: 191-92; Thuro 1976: 44). The patent date of 1883 probably refers to the innovative method of attaching the prongs, as stated in the patent "without rivets or solder." (a) Wick tube holder. (b) Ventilation plate, side view. (c) Ventilation plate, top view.

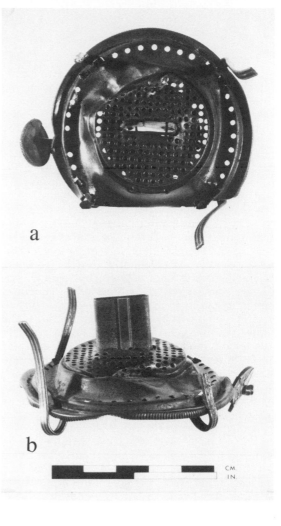

Figure 57. A kerosene burner with collar attached from the attic of the fur loft at Lower Fort Garry (1K99B1-44). The wick tube on this burner is 2.2 cm wide; the diameter of the collar is 3.1 cm. The collar is a style which pre-dates 1867. The chimney seating is the gallery type with a set-screw device for holding the base of the chimney. One of the three original set-screws is still in position. A loose deflector with a low dome is set within the gallery, upon which the base of the chimney rested. There are two marks on this burner; on the thumb wheel is the word NOVELTY and five stars. On the inside of the burner is the maker's mark, HOLMES, BOOTH AND ATWOOD MFG. CO. WATERBURY. CONN. This company existed only in the years between 1869 and 1871, which gives a narrow dating range for this particular burner (Russell 1968: 188). The unique feature to this burner, which qualifies it for the name "Novelty," is the device which allows the wick to be trimmed without removing the chimney or the deflector. The perforated cylindrical area beneath the ventilator plate can be pulled down away from the plate exposing the top end of the wick tube.

Figure 58. A kerosene burner from Fort St. James, British Columbia (3T19M2-2). This burner was from the area of the fill at the perimeter fence of the fort, dated to the late-19th century. The burner has a wick tube width of 1.6 cm and a collar diameter of 2.2 cm. The burner has a prong-type chimney seating. The maker is identified on the thumb wheel as the ONT.L.CO. This burner, based on the 1868 "Fireside" model, is typical of the standard flat wick burner found in North America during the late-19th and early-20th centuries. This particular burner is distinguished by the fact that it was manufactured by a Canadian firm, the Ontario Lantern Company of Hamilton, Ontario, which was established in 1892 (Russell 1968: 234-36). (a) Top view. (b) Side view.

Argand, in his experimentation with lighting, established the principle of enclosing the lamp flame within an open glass cylinder to create an artificial draft and improve the brightness of the light. His lamp was patented in England in 1784 (Russell 1968: 75-76), and glass lamp chimneys can date to any time shortly thereafter. On archaeological sites, however, lamp chimneys do not appear in significant quantities until after the widespread use of kerosene fuel and burners designed to be used with chimneys. Fragments of chimneys can be difficult to recognize as they often resemble tumblers and other drinking glasses or, occasionally, bottles. Lamp chimneys consist of an upper rim, a body, and a lower rim, each of which has sometimes been changed in shape to adapt the chimney to a change in another part of the lamp. The chimney's lower rim has a form appropriate to the chimney holder on the burner, the body of the chimney is modified to accommodate the flame which follows the shape of the wick and the fuel being burned, and the contour of the upper rim appears to be determined partly by glassmaking considerations and partly by style, depending on the period.

According to Russell (1968: 131, 135), kerosene had become the almost universal lamp fuel used in North America by 1864. By the late 1850s, kerosene was available in Montreal and Toronto and by 1860 in St. John's, Newfoundland. The new fuel was initially burned in lamps intended for other fuels, of which astral and solar lamps with Argand burners and chimneys probably gave the best results (Russell 1968: 141). Other types of lamps were converted to kerosene fuel by replacing the burner and adding a chimney. The adoption of the Vienna burner and its chimney for adapting lamps to kerosene introduced the bulbous-shaped chimney to North America. Vienna burners had a simple collar-like holder that required a chimney with a straight profile on the lower rim (Russell 1968: 142); this was probably inconvenient, as the diameter of the apertures of hand-made chimneys could not be relied upon to be of a standard size. A chimney holder that allowed for some variation in aperture diameter would have been desirable.

The Jones burner, patented in the United States in 1858 (Russell 1968: 150), had a coronet chimney holder with a screw of some kind, and chimneys to fit it had a flange at their base (Fig. 61). Later burners, developed during the 1860s, secured the chimney by pressure from inside and thus required a chimney with a straight lower part at least as high as the deflector, about 1 1/2 inches (Russell 1968: 182). The production of chimneys with the straight lower rim was reinforced with the introduction of the four-pronged chimney holder, after 1873 (Russell 1968: 225). Lower rims on lamp chimneys are usually more or less circular in horizontal cross section, but patents were granted for chimneys with oval and rectangular lower rims, made to fit oval or rectangular burners from ca. 1880 (Russell 1968: 226). These chimneys appear to have been mould blown.

The bulbous-bodied lamp chimney was used with burners that had a Liverpool button, patented in 1838 (Russell 1968: 85), although illustrations show a more angular bulb than the one used later on kerosene chimneys (see Knapp 1848: 484). The round-bulbed glass lamp chimney that continues in modern times was evidently in use in some European countries by the mid-1840s (Knapp 1848: 484), and came to replace almost completely the more angular version. An expanded area is apparently needed on a lamp chimney used with kerosene to correspond with the area of the flame, as the fuel and an improved draft produce a hot flame that could crack the glass of a narrow chimney. Although the bulbous part of the majority of kerosene lamp chimneys is circular in horizontal cross section, chimneys were also made with oval-shaped bulbs to accomodate the flame produced by a flat wick. The Ditheridge chimney patented in 1861 had a bulb with this shape and a circular top and bottom (Russell 1968: 183). Other oddly shaped chimneys include one based on an American patent of 1869, with an oval bulge, body, and top, and a cylindrical lower part (Russell 1968: 184).

Russell (1968: 282) characterizes the standard kerosene lamp chimney of the 19th century and later as having a straight, circular lower section intended to fit into prongs on the burner, an extended middle bulb, and a tapering top. The restricted neck and flaring upper rim are, apparently, features of the last 15 years of the 19th century. Russell (1968:

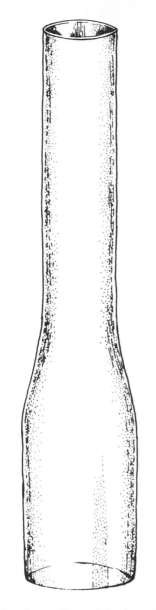

283) dates the decorated upper rim to 1885 and later in Canada; the popularity of decorated upper rims appears to begin earlier in the United States (Pyne Press 1972: 111). Acid-etched and painted decorations also begin to be popular in the late 1880s, and decoration on the tops and bodies of lamp chimneys of an earlier style, such as those with flaring lower rims, is possible beginning at that time.

Coloured lamp chimneys and shades were available in the 1840s but their use was limited as they tended to impart an unnatural colour to surrounding objects; ground glass (frosted) and milk glass shades and globes were a common means of both diffusing the light and deadening the effect of the flame (Knapp 1848: 156). Glass globes and shades were probably not common before the advent of kerosene and the development of cheap methods of decorating glass objects by such means as acid etching. Coloured glass lamp chimneys apparently had a period of use during the late-19th century (Russell 1968: 285), but they are not often found on historic sites in Canada.

Early in the 19th century gas was being used as a lighting fuel in factories, for street lighting, and in domestic buildings. Artificially manufactured illuminating gas was available in many cities and towns in Canada before kerosene was widely marketed (Russell 1968: 291). In its simplest form, the elements

Figure 59. Argand's original patented lamp relied on a chimney made of sheet metal which sat above the level of the flame; his first glass chimney was a straight cylinder so large in diameter that too much air was admitted (Knapp 1848: 142). Improved versions, such as the one illustrated here, were developed soon after Argand patented his lamp. They consist of a narrow cylinder with a deep constriction above the flame and narrower at the top than at the bottom. The constriction decreased the draft and directed it advantageously to improve the light (Knapp 1848: 142). This was the most familiar form of glass

chimney in use for lamps and gas lights in the pre-kerosene era (Knapp 1848: 461-62, 484) and was used on burners with flat, round, and semicircular wicks (Knapp 1848: 142). A revival of the Argand type of burner occurred during the 1870s in the form of student lamps and folded wick types (Russell 1968: 215) and the chimney was revived at the same time, although it probably continued in use all through the period. A chimney similar to that illustrated was recovered from a privy in Quebec City dating from the late-1820s to the early-1830s, and probably relates to the early patented lamp or to gas lighting. In later archaeological contexts, this shape would more likely be an indication of the renewed use of the burner type. Dominion Glass Company (post 1913: 52) offers a similar chimney in an early-20th century catalogue, and the American firm of MacBeth-Evans (Pyne Press 1972) manufactured it in ca. 1900. (Drawing by Susan Laurie-Bourque.)

Figure 61. Chimney with flanged lower rim. This example is modelled after an early-20th century Dominion Glass Company catalogue illustration (post 1913: 31) and is similar to one which Russell (1968: 151) calls characteristic of chimneys of the late-1850s and early-1860s, the early days of kerosene use. The vertical flange on the lower rim is intended to be held on the burner by a screw or a coronet, and is unsuitable for pronged holders and those using internal pressure to secure the chimney. A fragment of the flanged rim of such a chimney, recovered from a disturbed context at Fort George, is of colourless non-lead glass, and was presumed to be a bottle fragment. Chimneys of this type have a long period of use, although they should occur less frequently towards the end of the 19th century. (Drawing by Susan Laurie-Bourque.)

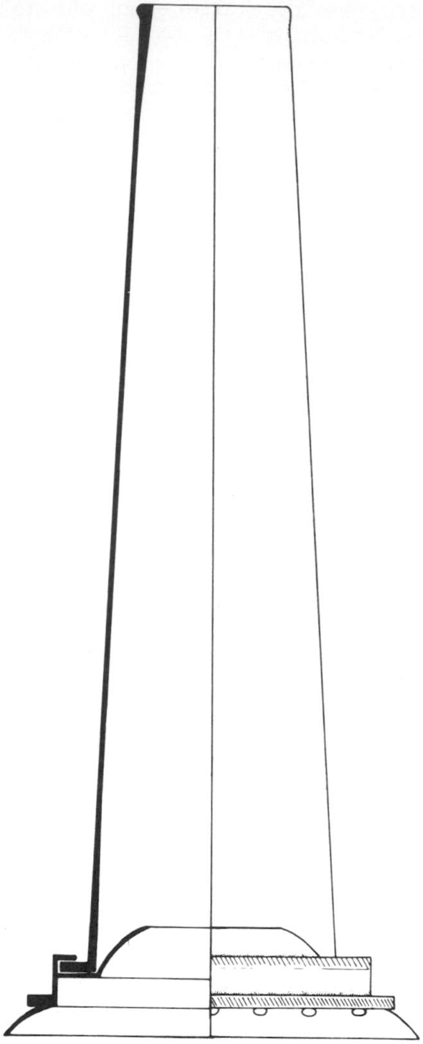

Figure 60. Solar lamp chimney. The solar lamp, based on Argand's burner, made its appearance during the 1840s (Russell 1968: 123-29). It could be used with whale or olive oil, but was particularly noted for its efficient use of lard fuel. The all-metal construction of the solar lamp transferred heat to the fuel to prevent its congealing. The flame produced by the solar burner was high, narrow, and very bright, and a high narrow chimney added force to the draft. This type of chimney is shown in an American catalogue of ca. 1850, reproduced in Russell (1968: 129), both on its own and in combination with a globe. In the latter case, the chimney protrudes far beyond the top of the globe. Three fragmentary examples of solar chimneys in our collection are from an

1838-50 context at Ft. Coteau-du-Lac, a disturbed area containing artifacts dating from 1865 to ca. 1870 at Ft. George, and in association with the hospital at Ft. Lennox, dating 1814-70. Both the lamp and its chimney appear to have gone out of use as kerosene replaced other types of lighting fuels. By the early-20th century, chimneys for solar lamps were not offered for sale at all in the catalogues of the Dominion Glass Company, Diamond Flint Glass Company, and MacBeth-Evans Glass Company. Russell (1968: 126, 129) has found a number of solar lamp bodies that have been converted to burn kerosene; many of these modifications appear to have been made by the manufacturers (see also Fig. 50). (Drawing by Susan Laurie-Bourque.)

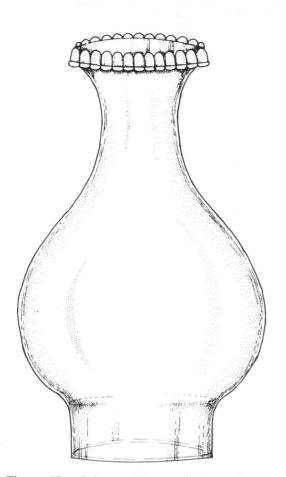

Figure 62. A lamp chimney (6G2B2-6) of the later years of the 19th century, from about 1885 and continuing into the 20th century (Russell 1968: 282). The exaggerated bulge, the restricted neck, and the decorated upper rim are all features of lamp chimneys of that period. In addition, this example is made of non-lead glass, probably based on Leighton's soda-lime formula developed in 1864 (McKearin and McKearin 1948: 8). This feature may indicate North American origin. Canadian and American manufacturers of lamp chimneys offer their goods in either lead or non-lead glass in the late-19th and early-20th centuries (Diamond Flint Glass Co. 1903-13: 3-4; Pyne Press 1972: 113). Decorated versions of this shape appear to have been very popular, and MacBeth-Evans (Pyne Press 1972: 116-17) manufactured its large bulge chimneys with etched, engraved, or painted designs, as well as plain. This style of chimney appears to predominate on archaeological sites of the late-19th and early-20th centuries. (Drawing by Susan Laurie-Bourque.)

required for gas lighting were a jet at one end of a pipe, a stop-cock at the wall end of the pipe, and a connection with the gas distributing system (Russell 1968: 293-96). Nozzles of different types produced flames with various shapes, and an Argand burner could be added for a more intense light (Knapp 1848: 184-85). In the late-19th century there were more elaborate versions available, and lamp shades could be used to dress up the outlet and to diffuse or direct the light. Many of these, shown in catalogues of the late-19th and early-20th centuries, were also advertised for use on electric lighting, the fixtures for which, at the beginning, were as crude as those for gas (see Lafferty 1969: 25-45). New developments in burners improved gas light, and portability was achieved by connecting the lamp to the gas intake by a long rubber hose. Although the quality of the light was controlled by adjusting the flow of the gas, it was found that a chimney steadied the flame by protecting it from drafts; an Argand burner used on a gas jet would not produce a smokeless flame without a chimney (Knapp 1848: 214). With the development of the Welsbach burner and mantle in 1885 (Myers 1978: 207), glass chimneys of a different shape came into use. The mantle is a cotton hood or sack, impregnated with chemicals, which is mounted over a burner and the fabric burned away. This leaves an ashy structure which the flame heats to incandescence to produce a bright white light (Russell 1968: 296-98). Incandescent gas lighting competed with electricity in the early period and was very popular in Britain (Myers 1978: 207) and other European countries, although as late as 1895 it was still a novelty in some places in Canada (Russell 1968: 297). The Welsbach mantle, with its narrow cylindrical shape, required a narrow cylindrical chimney to accentuate the draft (Russell 1968: 298). Commercial production of natural gas and the development of acetylene lighting in the late-19th century encouraged the move towards lighting with gas and apparently seriously threatened the development of electric lighting (see Myers 1978: 207).

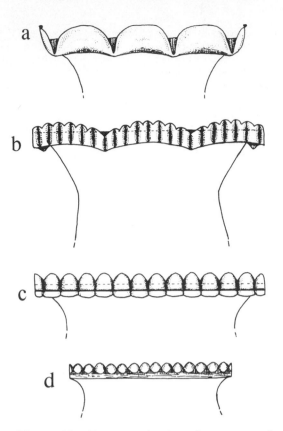

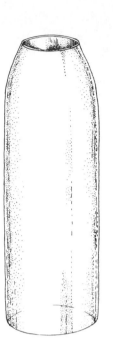

Figure 63. Decorated rim fragments from lamp chimneys. The decorations vary and the glass content in some cases is lead. The decorations on (a) and (b) appear to have been formed by hand-held tools; (c) and (d) are examples of decorations made with a template. Decorated chimney tops became popular during the 1870s in the United States (Pyne Press 1972: 111) — a crimping machine for producing a pie crust edging was patented there in 1877 and one for making a beaded decoration in 1883 (Pyne Press 1972: 111) — but Russell (1968: 225, 283) has found that decorated upper rims on chimneys are rare in Canada before ca. 1885. Canadian lamp chimney catalogues of the early-20th century (Dominion Glass Company, Diamond Flint Glass Company) show decorated rims on chimneys of the bulbous type, with narrowed neck and flaring top (see Fig. 62) and a lower rim that can be either flanged or straight. However, the majority of the chimneys in these catalogues have plain tops, possibly reflecting a trend towards decorated shades and globes at this period, as a decorated chimney top in combination with a globe appears to be rare. (Drawings by Susan Laurie-Bourque.)

Figure 64. This chimney is referred to as a "straight" in a Canadian glass catalogue of the early-20th century (Diamond Flint Glass Co. 1903-13: 16-17). There, it is a kerosene lamp chimney, intended for use in a hall lamp, and the manufacturer recommends that a different style of chimney with an extra slim bulb be used instead. Since the bulbous body on kerosene lamp chimneys reconciles a hot flame and glass in proximity, by implication, the straight shape is intended for use where dim lighting is required. Hall lamps are an enclosed, hanging lighting device, with font, burner, and chimney set inside an open-topped glass globe with draft holes at the base (Russell 1968: 223). They pre-date the kerosene era: Apsley Pellatt's English firm of Falcon Glass Works includes hall and passage lamps in a catalogue of ca. 1840, with other articles usually kept in stock (Wakefield 1968: 53). Ornate hanging lamps became popular in the last quarter of the 19th century and into the 20th century, all lamps among them, but whether their use continued uninterrupted or was revived at that time is not clear (Russell 1968: 223, 270). A fragment in partial lead glass from a chimney similar to this one was found in an English latrine at Signal Hill built after 1831, and was probably used with a lamp that burned vegetable or animal oil. (Drawing by Susan Laurie-Bourque.)

Figure 65. Incandescent gas chimneys. In early-20th century chimney catalogues, these objects are referred to as globes or cylinders, perhaps to distinguish them from kerosene chimneys, but they appear to perform the same function as kerosene chimneys and to be of similar size and glass thickness (Diamond Flint Glass Co. 1903-13: 73-79). The Phoenix Glass Company of Pennsylvania, in a ca. 1897 catalogue, shows these objects in combination with a dome-shaped glass shade, the two decorated by acid etching, sandblasting, flashing (or staining), in complementary designs. The chimney provides support for the shade, the latter resting by a constriction in its top on the expanded portion at the base of the neck of the chimney (Lafferty 1969: 61). The shape of this incandescent gas chimney differs significantly from those used for kerosene lamps, although the diameters of bottom and top and the overall height do not. Archaeologically retrieved fragments from undecorated versions are likely to resemble tumblers; the most diagnostic feature is the bump at the top of the bulb. Decoration, particularly if accompanied by a ground rim, can distinguish fragments of chimneys from other glassware forms. Chimneys of this shape should probably not be dated earlier than ca. 1890. (Drawings by Susan Laurie-Bourque.)

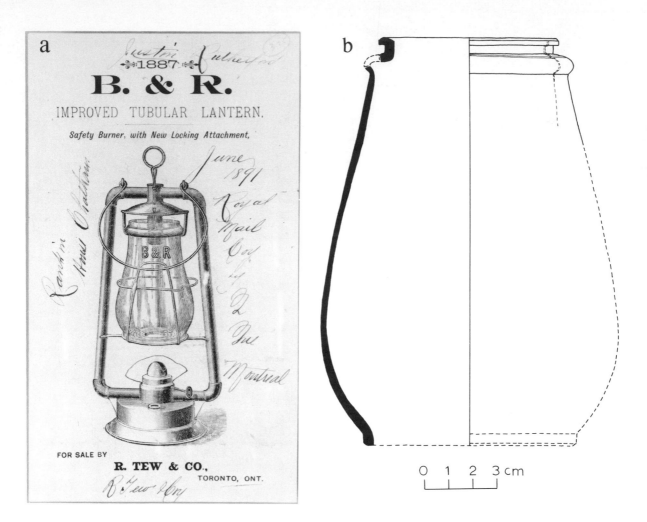

Figure 66. Lantern and chimney (1T1G6-1829). Glass chimneys for outdoor lighting seem to be distinguishable by their glass density, thicker than those used for indoor lamps, and by their method of manufacture; those in the Headquarters collection are ground at the top and bottom and have been contact mould blown (b). The chimney illustrated in the advertisement (a) has a moulded ridge near the top, intended to fit under a metal part of the lamp that raises and lowers the chimney. The same chimney is offered in a Canadian catalogue of the early-20th century (Diamond Flint Glass Co. 1903-13: 66-67). Lantern chimneys are shorter and sturdier than indoor chimneys, averaging less than 7 inches (17.8 cm) in height (Dominion Glass Company post 1913: 59-62; Diamond Flint Glass Co. 1903 13: 66-71). (Original owned by G.L. Miller; drawing by J. Moussette.)

LANTERNS

Lanterns are containers or enclosures for the protection of lighting devices. There are many variations, according to their use. Mainly intended for exterior use or in drafty locations, lanterns can be hand-held devices or mounted in permanent positions. Several specialized lantern types are designed for specific uses, e.g. on ships, trains, coaches, bicycles, etc. The more common forms are used for street, house, and hall lamps; portable versions are used in barns and out-buildings or for outdoor activities. Lantern forms have been devised to accommodate such different light sources as candles, liquid fuel lamps, and electric lights.

The sides of a lantern must be such that light can penetrate, yet be sufficiently dense so as to exclude drafts. Horn, mica, and glass are historically the materials used most extensively (Lindsay 1970: 51-53). Another solution, used in inexpensive all-metal lanterns, makes use of many small perforations in the sides of the lantern (Watkins 1966: 362).

When a burning fuel is used in a lantern, an air supply must be provided. Holes can be made at the base of the sides or in the bottom of the lantern to ensure an air flow to the flame. Another hole must be located at the top of the lantern to allow fumes and hot air to escape.

Lanterns Found in Archaeological Contexts

Examples of metal lanterns are rare in archaeological artifact collections, possibly because fragmentation has made identification of metal parts difficult. Lanterns were made in quantity in brass or tinplate in America. Brass objects are seldom found on historic archaeological sites because of the value of the metal as a raw material, and tinplate, as stated before, has a poor record of survival.

Glass lantern chimneys are more easily recognized, and may record the instance of lanterns more frequently than metal parts in archaeological contexts.

Figure 67. Lantern base from Fort Wellington, Ontario (2H39D14-4). This lantern was found on the exterior of the foundations of a house outside the fort near the St. Lawrence River believed to be a guardhouse dating to the early-19th century. The lantern was made of tinplate. The base was formed by stamping and has a wire-rolled edge. The lantern is cylindrical; the base is 14.7 cm in diameter. The interior of the lantern measures 9.5 cm in diameter. The inner base of the lantern is missing, so that the method of lighting cannot be determined. There is a row of holes at the lower part of the sides, suggesting that the lantern had originally contained a fuel light requiring an air flow, such as a candle or a lamp. The design of this lantern is that of a hand-held device with a base for steadying it when set down.

ACETYLENE LAMPS

Acetylene is a gas produced by the action of water on calcium carbide; the calcium carbide combines with water to form acetylene and slaked lime. Because acetylene produces a brilliant white flame when burned in the presence of a sufficient amount of air, it was used in several lighting devices.

In the late 1890s the use of acetylene for lighting public places was popular. However it was not well suited for home lighting because of its unpleasant characteristics, in particular the bad smell and corrosiveness of the carbide (Russell 1968: 300).

This gas also became very popular for use as fuel for lights on vehicles, bicycles and cars, and for miners' lamps. It is still in current use in spelunkers' head lamps.

The lamp described in Fig. 68 is believed to be a model for a bicycle. It consists of a water tank attached to a hermetically fastened burner on the acetylene generator. The water tank, located level with the flame, included a reflector which is now missing. The lamp could be directed vertically by adjusting the clamp mounting.

The tank was filled regularly by means of a pipe, perhaps originally equipped with a small funnel. The water flowed down the pipe by gravity and dripped into the carbide reservoir. A stopcock regulated the flow and thereby the amount of gas produced. The acetylene produced in the generator rose to the burner, the nozzle of which is missing here, along a coil, part of which passed through the cover.

Almost all the metal parts of the lamp are of bronze or copper. The outside is plated.

On the back of the water tank, in stamped letters, is the manufacturer's mark THE BADGER BRASS MFG. CO. KENOSHA WIS. U S A. The model is SCLAR (or SOLAR). Underneath the bottom of the generator are the dates of the patents:

PAT. IN U.S.
    FEB. 4. 1896
    JAN. 31. 1899
    FEB. 7. 1899
PAT. IN ENGLAND
    FEB. 4. 1896
    MAY. 12. 1896
PAT. IN CANADA
    JUN. 15. 1899.

The item comes from a salvage excavation in the Richelieu River at Saint-Jean à Cantic, Quebec. The carbide reservoir is still full of hardened lime. The pressure plate designed to prevent movement of the carbide, as well as part of the coil that acted as a spring on it, are still in place.

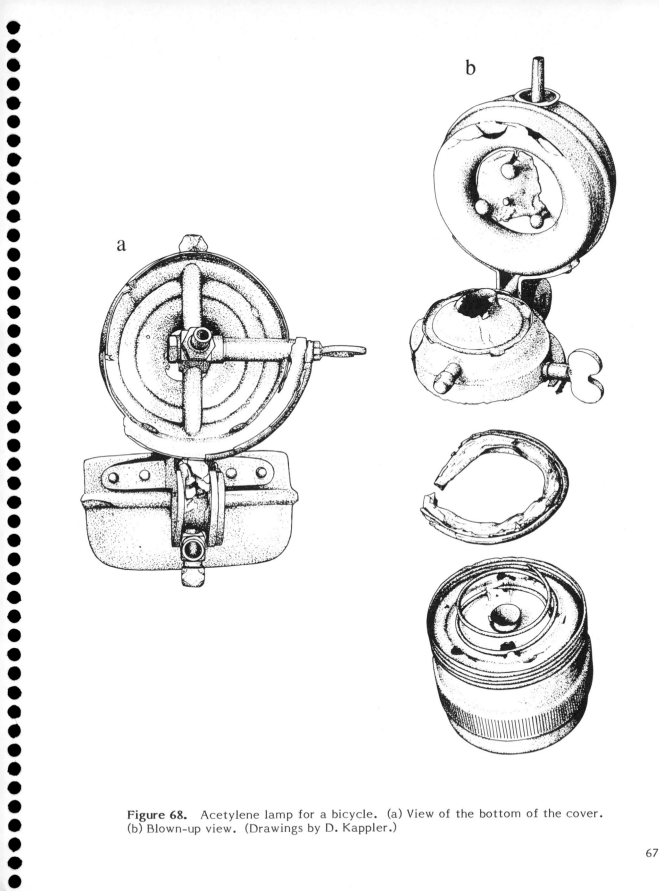

Figure 68. Acetylene lamp for a bicycle. (a) View of the bottom of the cover. (b) Blown-up view. (Drawings by D. Kappler.)

Unfortunately this discussion on the methods of electric lighting is not based on an archaeological collection. Most of the pieces described and illustrated here are gifts to the research collection from headquarters in Ottawa. However bulbs and sockets are frequently found in the material from various sites excavated by our regional offices and other groups.

The small collection gathered here shows briefly the general evolution and variety of electrical lighting equipment. We define the major stylistic and technical characteristics that make classification, systematic description, and dating of these artifacts possible.

Because electrical lights underwent numerous and varied changes in a short time period, their remnants, especially fragments of lamps of all kinds, seem to be an ideal way of dating sites. The incandescent electric bulb, for example, which makes up the largest proportion of artifacts related to electric lighting, was developed over a period of approximately 100 years (1840-1940). During this century it was improved by dozens of innovations, most of which were patented and advertised during a certain period of time.

Unfortunately for the archaeologist, and even for the simple collector of bulbs and insulators, delays in distribution and sometimes the lack of distribution make it very difficult to assign specific dates, especially in Canada. All these improvements were made by English or American manufacturers. The user often did not get new models until several years later; their distribution was relatively slow and was subject to geographic factors and urban development. Several models of lamps never arrived here; others stayed longer than elsewhere.

For a long time electrical lights were a commodity reserved for industry, then for the inhabitants of the larger urban centres. Therefore the dates in this guide should be used solely as the dates the lights were invented in England or the United States, and Canadian industrial development must be taken into account to determine a more specific date for a site. As we shall see briefly in the history, electrical lights followed the installation of an electricity network, which in turn depended on the modernization of the manufacturing industries that had the most to

gain from it. In addition it represents only one aspect of the use of electric power in industries and urban centres, and should be studied at the same time as the development of transportation, the use of stationary motive power, and improved communications.

Some Historic Facts: Lighting and Motive Power in Canada

Electricity for lighting streets and public places came into general use at the end of the 19th century, first in countries where industrialization grew the fastest, especially England and the United States. Its subsequent expansion followed the development of electricity distribution networks, from the large cities along the United States border toward the country.

The arrival of electric power for distribution was accompanied by profound technological changes in all areas of industrial activity; it made adjustments in numerous aspects of people's daily lives necessary; and it made possible or brought about the development of an incredible number of inventions in all scientific areas.

This new form of energy spread relatively fast because its advantages over steam as the motive power in heavy industries such as mining and textiles were discovered very quickly. Electricity also made new methods of lighting possible that were more flexible, less dangerous, and more suited to the uses; it also offered the possibility of making never-ending improvements to means of communication.

In Canada, urban electrical lights developed quickly during the last two decades of the 19th century and early-20th century; the first hydroelectric power plants were built (the line between Niagara and Hamilton was finished in 1890) and power distribution companies multiplied by mains. For example:

Canada. - In his annual summary of the progress of electricity support in Canada, Mr. Geo. Johnson states that the number of electric lighting companies has increased from 259 in 1898 and 306 in 1901 to 312 in 1902. Arc lights increased from 10,389 in 1898 to 12,884 in 1902, and incandescent lamps number 995,056, an increase of 179,380 over 1901 and 531,441 over 1898. Reckoning each arc equal to 10 incandescents, the use of electricity as a light giver has developed from 565,505 lamps in 1898 and 943,676 in 1901 to 1,123,896 in 1902. Of the 312 electrical companies Ontario has 195 *(The Electrician, Vol. 50, No. 12, 9 January 1903, p. 489)*.

In southern Ontario and Quebec electricity was adopted as a source of energy for lighting and motors by the large mining companies and the pulp and paper industry. Its production was made easier by the volume of water in the St. Lawrence River basin, which made it possible to build power plants along the river and several of its tributaries.

Nevertheless electricity had to coexist for a long time with the more traditional forms of energy of the industrial revolution which, in some aspects, still remained more advantageous in the Canadian context. Early in this century industry still preferred hydraulic power for providing motive power, although electricity was seen as an ideal alternative.

Electricity was, however, beginning to modernize the power situation. Such modernization was slow and as late as 1925 hydraulic turbines and water wheels provided more mechanical power than steam engines in the woollen cloth industry. By that date electricity supplied well over twice as much as either of the above sources, whereas in 1911 water power supplied two and a half times as much mechanical power as electricity.

••
The trend in energy utilization in southern Ontario was from a predominance of water power in 1871 through a period of steam power until the first decade of the 20th century to a strong reliance on electricity in 1921 (Walker and Bater 1974: 64-65).

Because of supply difficulties in Canada, coal had to be imported. It was favoured to a certain extent for heating and the production of steam until the middle of the century.

Nevertheless, the adequacy of hydroelectricity as an energy base for large scale industrialization must not be exaggerated. Hydro power has been a *sine qua non* for the industrial development which has in fact taken place in central Canada, but, as we have already pointed out, imported coal has been no less indispensable. Of the total energy provided in Canada by water power and mineral fuels together, coal has in recent years provided more than one-half. In Ontario and Quebec, imported coal furnished in 1943 over 50 per cent of the total energy consumed.

••
Electricity is a convenient and efficient form of energy for motive power - over 80 per cent of the power equipment installed in the mining and manufacturing industries of Canada is electrically driven - but as a source of industrial heat it is neither as efficient in the technical sense nor generally as economical as direct combustion of fuels (Easterbrook and Aitken 1958: 527).

It seems that the electrical lighting of streets and homes was clearly linked to the adoption of electricity as the motive power for industry. Lighting alone, installed in an area the least bit isolated from manufacturing centres and the axes connecting them with the St. Lawrence and Great Lakes basins, would never become profitable. This explains the speed, and in some cases the slowness, with which electricity came into general use in the cities and the countryside during the first half of the century.

In Canada, as in other industrialized countries, the study of the installation of an electrical network is based on the history of the producing companies and of the major industries. All the regions of the country are not equally endowed with the availability of hydroelectric resources. Quebec and Ontario, privileged from this point of view, were able to set up during the first half of this century a production network that was powerful enough to provide for most of their energy needs.

As an indication, based on a few very general statistics, the following describes the situation across the country around the end of the first half of this century.

Ontario and Quebec between them possess more than 80 per cent of Canada's developed hydro-electric capacity primarily because they have been able to draw on the water-power of the Precambrian Shield and the St. Lawrence

River. Cheap hydro-electric power, together with the mineral and forest resources of the Shield, has been the principal factor responsible for the development in these provinces of the pulp and paper industry and the non-ferrous metal smelting and refining industry, two of the largest industrial power-consumers. The main power sites are at Niagara Falls and on the rivers of the St. lawrence drainage system, particularly the St. Lawrence itself, the St. Maurice, the Saguenay, and the Ottawa and its tributaries; ... There are also large installations on the rivers draining into Lake Superior and James Bay, which supply the mining districts of the northern part of the provinces. In Ontario, largely because the original development was based in the exploitation of a single very large source of power — Niagara Falls — generation and transmission are in the hands of a public body, the Hydro-Electric Power Commission in Ontario. In Quebec the industry grew up under private enterprise.

In none of the other provinces except British Columbia and possibly Newfoundland has hydro-electricity played as large a role in industrial development as it has in Ontario and Quebec.

In British Columbia the impact of hydro-electric power has been no less remarkable than in central Canada. The total waterpower resources of the area were estimated in 1940 at 5.2 million horsepower; of this total no more than a fraction has been harnessed for the production of electricity. The principal industrial consumers are, as in central Canada, the pulp-and-paper, light metals, and chemical industries. ... The most important installations are on the Kootenay River near Nelson, on the State River near Vancouver, and on the North Arm of Burrard Inlet. ... Water-power supplied 37.9 per cent of the total energy consumption of the region in 1943, as compared with 32.7 per cent from coal and 29.4 from petroleum (Easterbrook and Aitken 1958: 525-26).

Elsewhere, the production of electric power is hindered by limited hydroelectric resources and the industry must use thermal power stations that operate on organic fuels.

In the Maritimes there are developed power sites on the Mersey River in Nova Scotia and on the St. John River in New Brunswick. The total hydro-power resources of the region, however, are relatively small, amounting to only about 500,000 horsepower, while the availability of locally produced coal gives an advantage to steam-generated electricity. Of the total energy consumption of the Maritimes in 1943, coal (either burned directly or in thermal electricity plants) provided almost 70 per cent and water-power only 8.5 per cent; this compares with 52.6 per cent for coal and 37.8 per cent for water power in Ontario and Quebec. Pulp-and-paper mills at Liverpool, N.S., and at Edmundston and Dalhousie, N.B., are, however dependent on hydro-electricity.

In the prairie provinces, too, water-power resources are relatively limited, being estimated at not more than 1,800,000 horsepower. The principal power sites are located at the north-eastern margin of the prairie region, where it adjoins the Precambrian Shield, and on the western margin, near the eastern range of the Rockies. There are important hydro-electric stations on the Bow River and the Winnipeg River in Manitoba, ... Naturally gas and petroleum together furnish just over 30 per cent of the total energy requirements of this region, coal 58.7 per cent, and water-power only 11 per cent (Easterbrook and Aitken 1958: 526).

Electrical Lighting Equipment

Historically, there are three ways of producing light using electricity:
1. By heating a conducting body to incandescence. This is the phenomenon that takes place in an electric light bulb where the filament is heated to white-hot when a current of sufficient intensity passes through it.
2. By heating two graphite electrodes and ionizing the air between them. This is the arc lamp, in frequent use until the middle of the 20th century for street lighting in large cities. Two graphite rods are brought into contact with sufficient potential between them. When a spark is produced, they are gradually moved a few milli-

metres apart and a brilliant luminous arc will continue to be produced. When the two elecrodes are first placed in contact, the strong resistance between them causes incandescence at their ends. After they are separated, the highly ionized air becomes a conductor and thus enables the electrons to continue to pass through and produce a luminous plasma.

3. By ionization of a rarified gas or vapour contained in a transparent tube or bulb. When the current goes through the gas, in certain conditions, some electrons from its atoms pass from one energy level to another by freeing photons. This principle, put into application relatively recently, is the one used in fluorescent tubes, neon lighting for advertising, and the flash-type discharge lamps used in photography and signaling.

INCANDESCENT LAMP

The incandescent lamp is the kind of electric lighting found the most frequently in our archaeological sites from the post-industrial period. Because of its simplicity, its relatively low cost, and the east with which it could be used in a large number of places, this type of lighting was adopted quickly and became the type of lighting most commonly used in most areas of industrial and domestic life during the early decades of this century.

A basic lamp will include the following parts (Fig. 69):

1. One or more filaments.
2. A glass bulb, which generally contains a partial vacuum to protect the incandescent filament from oxidizing too rapidly.
3. The foot of the lamp, which is a kind of glass tube or node located at the base of the bulb to hold the filament supports or, in the case of older lamps, the filament itself.
4. The wire supports to hold and immobilize the filament.
5. The pumping opening, through which the air is removed from the bulb. After being soldered with glass, it is shaped like a small pin at the top of the bulb, or at the bottom in the base.
6. The base, which generally consists of a metallic band around the bottom of the bulb. It of course serves to fasten the lamp into its socket, but also to provide contact with one side of the line. Bases are generally of the bayonet or screw type (Edison type).
7. The base contact, a small metallic disk or point at the bottom of the base that provides the connection with the other side of the line.
8. The gas. The nature and pressure of the gas inside the bulb can vary; it often consists of rarified air or, in more modern lamps, a rare gas such as argon or krypton under low pressure.

To make the analysis, description, and approximate dating of lamps easier for researchers, we will give a brief description of the evolution of the incandescent lamp, noting for each step in its development the most important technical characteristic(s) to remember.

The evolution of the incandescent lamp in fact follows its performance. The inventor's problem always consisted of increasing the brightness of the luminous flux produced by the filament while increasing the life span of the filament.

Between 1840 and 1879 a series of experiments resulted in the creation of rudimentary lamps that could not be marketed because of their low intensity and extremely short life span. Very few of these lamps were manufactured and the ones that have survived to the present day are found, for the most part, in museums of technology. Nevertheless they played an important role in that they enabled scientists of the day to experiment with filaments made from a wide variety of substances and thus prepared the way for the success of people like Edison. The main accomplishments of this period were as follows.

In 1840 the Moleyns of Cheltenham obtained the first patent for a filament lamp. That same year Grove said he could read by the light of a lamp he had made using a platinum filament covered with a drinking glass turned upside down in a plate of water, with energy supplied by Grove or Bunsen batteries (O'Dea 1958: 11).

In 1845 Starr of Cincinnati obtained a patent in England for a lamp with a carbon filament enclosed in a vacuum.

Between 1848 and 1878 experiments were conducted by Joseph William Swan, working from Starr's and Staite's notes. These produced the first truly functional lamp about the same time as Edison. Swan demonstrated this lamp in 1878, whereas Edison had his

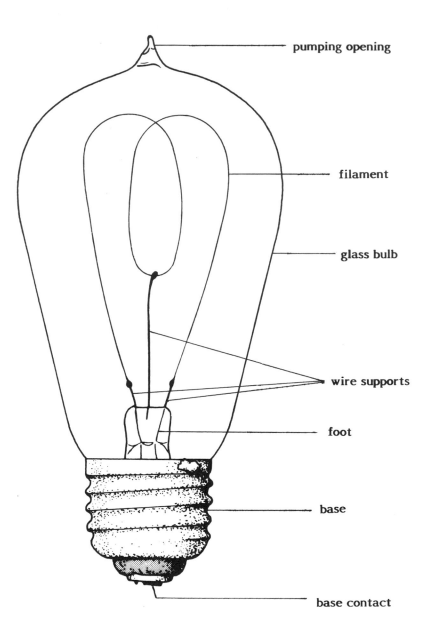

pumping opening

filament

glass bulb

wire supports

foot

base

base contact

Figure 69. Incandescent lamp and its parts. (Drawing by D. Kappler.)

patented around the end of 1879 (O'Dea 1958: 11).

Edison himself experimented a great deal to find a more durable and brighter filament. He first tried platinum, then an alloy of platinum and iridium in a vacuum bulb. These filaments, which had to be brought to a temperature near their fusion point, broke quickly and the lamps did not last a long time.

Tests with carbon-based filaments followed—a mixture of tar and soot, and finally in October 1879, the year that the great invention was patented, the carbonized thread, which gave the best results obtained until then.

Research continued in the Menlo Park laboratory in New Jersey; after 1880 Edison experimented with carbonized paper and carbonized bamboo fibres. The lamps with bamboo filaments were superior to all their predecessors and were used until 1898 (Cox 1979: 46).

In 1881 the Edison lamp was equipped with a screw base, which was to become one of its distinctive characteristics.

At the same time as the incandescent lamp was being marketed, thanks to Edison's efforts, a large number of accessories and equipment needed to operate and maintain an electric circuit, such as dynamos for power, meters to measure the current, sockets, switches, and fuses (Cox 1979: 47), and even independent generating sets, were being marketed and mass produced.

From then until around 1940 improvements to the filament lamp continued with numerous innovations, which, among other things, increased its life span from a few to more than 1000 hours. During this time the cost of filament lamps continued to decrease.

We do not intend to describe in detail here all these improvements with their national variations. Instead we will briefly point out the ones that are the most significant and the most easily noted by archaeologists and collectors.

In 1894 Lord Rayleigh invented a filament lamp that worked in argon. The filament survived for a longer period of time in this gas than in rarified air. This was the forerunner of the rare gas lamps that are in general use today.

In 1897 Dr. Nernst invented a type of lamp with multiple filaments of thorium, cerium, and so forth; these substances were used at the time in the manufacture of incandescent mantles for gas lamps and could be made

conductors by the effect of a heating element passing through their centres. These lamps did not require a vacuum for their operation, and they were very popular during a period of approximately 10 years O'Dea 1958: 14). Early in this century and certainly until the 1920s numerous photometric tests were done on the Nernst lamp to improve its performance.

In 1898 an injected carbon filament was developed which was to replace the carbonized bamboo filaments used in the Edison lamps. In this new procedure cotton was dissolved in a solution, then injected and compressed in a mould to harden it. The filament was then carbonized with heat before being used in the lamp. This type of filament was replaced with the tantalum and tungsten filament around 1910 (Cox 1979: 46-47).

In 1897 or 1903 the osmium filament was created; with this metal a brighter light could be attained with the same consumption of electricity. But osmium was rare, and these lamps were quickly abandoned (O'Dea 1958: 14; Cox 1979: 47).

In 1905 Von Bolten and Feuerlein produced the drawn tantalum filament. Although tantalum did not break as easily as osmium, it had a tendency to soften under the effect of heat and the filament had to be fastened to the bulb to resist strong vibrations when the lamp was lit (O'Dea 1958: 14).

After 1907 the commercial production of tungsten filaments in the United States was one of the most important milestones in the development of the filament lamp, and at the same time marked the beginning of the era of the modern incandescent lamp (Cox 1979: 47) as we know it today. In fact a filament of tungsten, a metal with one of the highest fusion points, heated in a rare gas such as argon does not burn. This greatly increased its life span while preventing excessive blackening of the interior of the bulb. It should also be noted that tantalum-tungsten alloy filaments quickly replaced injected carbon filaments around 1910. Inside the bulb the filament was in the shape of a wire cage around a glass tube.

The year 1913 marked the beginning of the modern rare gas lamp. Langmuir developed a spiral filament obtained by winding a wire around a larger wire that was later dissolved in acid (O'Dea 1958: 14). The filament was now shaped more or less like a ring and was held in place with metallic wires anchored to the central glass tube. The bulb itself also underwent modifications. Its shape became rounder and acquired the pear-like profile that we know today. The development of glass-blowing and vacuum machines made it possible to pump and introduce gas through the base. The pumping opening disappeared from the top of the bulb and was now placed in the bottom part, in the base.

In 1925 another improvement to the bulb made its appearance even closer to that of present-day bulbs. Until then bulbs had been left clear or frosted on the outside, but now they were given a satiny interior which had the advantage of preventing glare from the bright incandescent filament and dispersing the light better.

From 1934 to 1936, depending on location, slight modifications in the shape of the filament gave it the familiar appearance of lamps today—the double spiral (O'Dea 1958: 14; Dubuisson 1968: 871). The wire is wound once into a spiral and then coiled again into a second spiral with a larger diameter.

These few notes on the morphology of the bulb and filament demonstrate that the date of a type of incandescent lamp can be determined quite accurately provided reasonably complete remnants are available. An analysis of the composition of the filament is a particularly useful way to determine the chronology of the various models. More accurate dates, especially for the numerous variants not described on these pages, can be determined only by using manufacturers' old catalogues or by directly contacting their public relations services.

Old bulbs were blown with blowpipes. The first attempt took place in Corning, New York, in 1870 (Davis 1949: 231). After the 1890s they were blown from glass tubes, and around 1894-95 their production was partially automated with blowing machines of the Owens type (Scoville 1948: 331). In 1927 the procedure was entirely automated (Davis 1949: 233).

ARC LAMP

The basic design of the arc lamp is very simple—two graphite electrodes held in place by a support and two conductors connected to a source with enough current to make it operate. However most of these lamps are complex machines, often including several dozen parts needed to start and operate the devices under various voltages. Nearly all include a system for regulating the resistance voltage to control the intensity of the current circulating through the lamp, and a solenoid which ensures the correct spacing between electrodes both when the lamp is lit and as the electrodes are consumed.

Following is a list of the parts that may be included in the construction of a sophisticated lamp dating from the early years of this century (Fig. 70)—a metallic lamp body to hold the mechanism, one or two glass bulbs, two graphite electrodes, graphite solenoid(s) (hollow coils forming the electromagnet) with their system of linkages consisting of levers, arms, screws, rivets, and springs, bits of wire, charge resistors (generally coils wound around porcelain insulators) or rheostats (variable resistance with a sliding coil), and various types of insulators (mica, asbestos, or porcelain).

When the lamp is not lit the ends of the two electrodes are in contact (Fig. 70B, C). When they are pulled apart a spark jumps between them and the regulated current, approximately 100 volts that also flows through the solenoid (Fig. 70E, F), activates the solenoid which pulls the top electrode a certain distance from the bottom one. The lamp is lit. The solenoid keeps this distance constant as the electrodes are consumed and adjusts it according to the voltage of the power supply.

The electric arc gave off much more light than the best contemporary incandescent lamps. A lamp of the type described here could burn for 125 hours without having to have the electrodes replaced.

It is impossible to establish the chronology of the development of the arc lamp as clearly as we did for the filament lamp. Its complexity, the large number of models, and the numerous improvements made to all the parts by a large number of manufacturers makes its exact dating rather difficult. In addition, we have only a few fragments of electrodes in our collections. Nearly complete arc lamps have not been found in archaeological excavations in Canada. When these machines broke down they could be repaired, unlike lamps that worked on other principles, and when public utilities decided to replace them they were often sold in lots to scrap merchants. Nevertheless some models can be dated on the basis of trademarks, patent numbers, and by referring to catalogue illustrations. Several types, including the one shown in Figure 70, are described and dated in various issues of *The Electrician*.

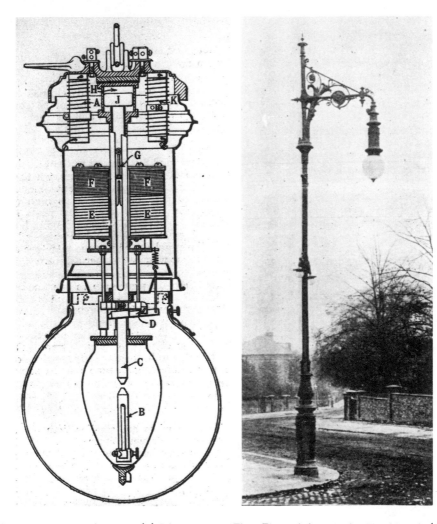

Figure 70. Arc lamp. (a) Diagram. *(The Electrician,* Vol. 54, No. 3 (1904), p. 100.)* (b) Photograph. *(The Electrician,* Vol. 52, No. 3 (1903), p. 88.)*

The first historical evidence of the creation of a luminous arc using electricity seems to be Watson's experiment in 1751 in which he created an arc inside a mercury barometer (O'Dea 1958: 11).

In 1810 Humphry Davy, inventor of the safety lamp for miners, demonstrated his carbon arc lamp at the *Royal Institution,* but the lack of dynamos and alternators to produce enough electricity prevented it from being put into general use at this time (O'Dea 1958: 11).

In 1857 Holmes built a continuous current generator capable of supplying power to the arc lamp and developed the first automatically regulated lamp (O'Dea 1958: 36). The Holmes

lamp and its generating set, driven by a steam engine, were installed in lighthouses shortly thereafter.

This lamp's most popular use was for lighting streets and public areas in large cities after about 1877 and until the 1950s, when it was replaced by high-intensity incandescent lamps.

During the first decades of the 20th century the arc lamp underwent many technical improvements to decrease its consumption of electricity and make its maintenance easier. Many new models were also put on the market.

For a long time the electric arc coexisted with the filament lamp, each having its own specific uses. The incandescent lamp has always been preferred for domestic lighting; the arc was not well adapted to this because of its highly intense and harsh light, its heat, odour, smoke, noise, and twinkling. Nevertheless, early in this century there were a few "miniature" models of arc lamps, mounted on a foot, for use in the home.

In the last 30 years the electric arc has gradually disappeared in Canada as a means of general lighting. It has survived for only a few special technical uses. It is still used as a light source in certain spectrometers and for movie projectors.

IONIZATION LAMP

The ionization lamp, also called a discharge tube or lamp, is the most recently invented method of electric lighting in general use. It has become a strong competitor of the filament lamp and its use in all spheres of human activity is growing. Its popularity stems from its advantages which are very definite and desirable in the context of modern life—bright light, low energy consumption, low cost, long life span, great adaptability to specialized uses, no maintenance, and a better spectral quality of light.

We do not have archaeological specimens of the ionization lamp although they have been found in excavation sites or parts of sites of recent occupation. The examples shown here illustrate the principle on which they operate and their variety, giving researchers a means of identifying, describing, and dating them.

In an ionization lamp, which is often in the shape of a tube or a large ball, the current passes from one metallic electrode to another through a gas. The free electrons displace certain electrons in the gas atoms, which are then said to be ionized. The electrons in the atoms move to a different energy "level" or "layer" by freeing photons — this is the direct production of light. In other cases the free electrons move toward the inner surface of the tube where they can produce light by the fluorescence of a thin layer of coating.

The shape and method of construction of ionization lamps may vary greatly, depending on the type of excitation, the nature of the gas, and the intended use of the lamp. The following is a simple classification (O'Dea 1958: 16) modified by us to meet the objectives of this project.

Types of Ionization Lamps

A) Non-fluorescent
  1. Cold cathode (all at low pressure) such as the Moore carbonic gas tube (1895); the Cooper-Hewitt mercury vapour tube (1900); the neon tube (1922).
  2. Hot cathode: (a) low pressure such as the Cooper-Hewitt mercury vapour lamp (1900) and the neon tube (1922); (b) high pressure such as the mercury vapour lamp (1932).

B) Fluorescent (all hot cathode)
  1. High pressure such as the mercury vapour lamp (1939, 1957).
  2. Low pressure such as mercury vapour fluorescent tubes (1940, 1957, 1948) (Fig. 71).

The cathode is the electrode that emits electrons in a vacuum tube. It is said to be hot when it includes an element that is heated to incandescence by an auxiliary current to promote the emission of electrons. The cold cathode does not have such a device. The pressure indicates the amount of gas in the tube. In a low-pressure lamp the gas pressure may be only 1/100th of an atmosphere whereas it may be several atmospheres in high-pressure lamps.

Ionization lamps, depending on their type, may include the following widely varying parts, mostly of metal or glass: one or more gas or vacuum bulbs, glass tubes sometimes in the shape of a U or a spiral, electrodes (rods, wires, disks, or grids), a glass, porcelain, or plastic base, terminals or contacts, and insulating material.

The first experiments with electrical discharges in rarified air date back as far as the 18th century. In 1709 Hawksbee, an Englishman, used an electrostatic machine to produce a discharge in a tube containing a partial vacuum (O'Dea 1958: 11).

In 1768 Canton obtained a luminescent compound with a calcium sulphite base by heating and mixing powdered oyster shell with sulphur (O'Dea 1958: 21).

However it was not until 1895 that the first truly functional apparatus was made—Moore's carbonic gas tube (O'Dea 1958: 21).

For documentary purposes we will briefly describe the main types of discharge lamps, the ones most widely used that are most likely to be encountered in archaeological sites.

Neon Tube for Promotional Lighting - It was invented by Georges Claude around 1810, but did not come into use for commercial advertising until 1922. The neon placed in the tube produced a red light. It is a low-pressure, cold-cathode lamp, generally operating under high voltage.

Modern Mercury Vapour Lamp - It was put on the market around 1932, in particular for lighting roads and public places. It was generally a high-pressure, hot-cathode lamp that produced an intense bluish light. The mercury lamp produced an arc inside a bulb that was itself enclosed in another bulb to retain as much of the heat that was needed for it to work as possible. After 1936 quartz bulbs were used to resist higher temperatures and pressure (O'Dea 1958: 23; Dubuisson 1968: 876).

Sodium Vapour Lamp - It operates under low pressure and has a cold cathode. Its introduction on the market also dates from 1932. This type of lighting has undergone improvements and gained in popularity for lighting roads and public monuments. It gives off a characteristic bright yellow light. It also has a double bulb. The tube where the discharge takes place is rather long and folded over, and the glass is specially made to resist the corrosive action of the sodium and the intense heat (O'Dea 1958: 23; Dubuisson 1968: 876).

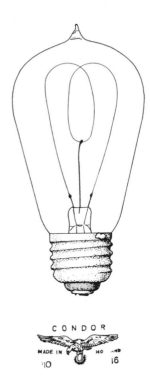

Figure 72. Incandescent lamp with its factory mark (1U2A1-82). Height: 11.5 cm. There is a single double-looped filament of unidentified composition. Note the three wire supports directly attached to the bulb. There is a pumping opening at the top, the bulb is clear, and the screw-type base is copper. Manufacturer: Condor, Netherlands. Approximate date: very early 20th century. General usage lamp impossible to specify.

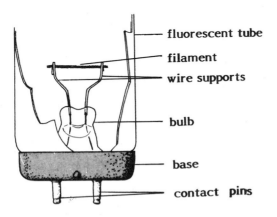

Figure 71. Fluorescent tube and its parts. (Drawing by D. Kappler.)

Fluorescent Tube (Fig. 71) - This lamp, in widespread use today for domestic as well as commercial lighting, gives off light through the fluorescence of a coating, generally a white powder, that covers its inside surface. Their use on a large scale began around the end of 1927. By late 1941 fluorescent tubes were available in North America in lengths varying from 9 inches to 5 feet (O'Dea 1958: 29). The modern tubes have low pressure and a hot cathode. They are easily recognized by their white interior coating. Their starting and operation also require the use of a transformer and starter often called a ballast (Dubuisson 1968: 872-73).

Figure 73. Miniature incandescent lamp (1U2A1-318). Height: 7.2 cm. The single multiple-looped filament is of unidentified composition. There is a pumping opening at the top, the bulb is clear, and the screw-type base is copper. Approximate date: very late 19th century or very early 20th century. Small lamps of this type were often made to be mounted in series (garlands or borders) for decoration or advertising.

a 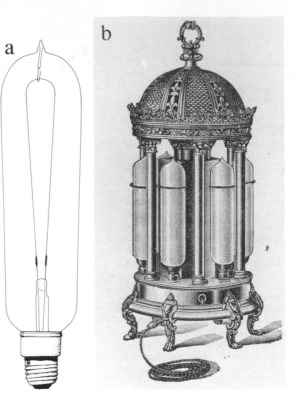 b

Figure 74. (a) Incandescent radiator lamp (1U2A1-297). Length: 29.4 cm. The single large-diameter filament is bent into a U and attached at the top to a hook anchored in the soldering of the pumping opening. The bulb is frosted on the outside. There is a copper screw-type cap. Manufacturer: General Electric. Approximate date: early 20th century, before 1925. This lamp was primarily designed to produce heat, as may be noted from its large filament. With several others mounted vertically, it constituted a source of heat for a small domestic heater (b) of a type in widespread use early in this century. (Drawing by D. Kappler; photograph from *The Electrician,* Vol. 50, No. 1 (1902), p. 37.)

Xenon Lamp - This is a cold-cathode tube in which an arc jumps through xenon. It is a very recent design. The light is a brilliant white, similar to daylight. Physically the bulb may be in various shapes, depending on the intended use. It is used especially for stage lighting in the theatre, for television, movies, and photography (Dubuisson 1968: 877). Instant discharge lamps of the flash type used in stroboscopes and for signaling are also xenon lamps.

To Describe Electrical Lights

Electric lighting equipment must be described in a rather complete and uniform manner so that it can be used in archaeological interpretation and the dating of sites. Below are listed the characteristics to be noted wherever possible when classifying lighting equipment.

NAME AND OPERATING PRINCIPLE

1. Incandescent filament lamps.
2. Carbon arc lamps.
3. Ionization lamps.

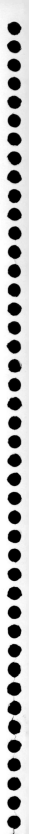

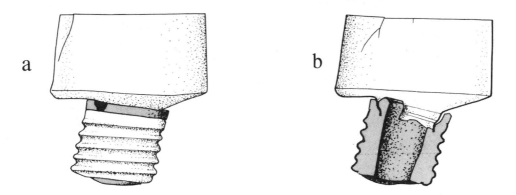

a   b

Figure 75. Copper base of a radiator lamp (1K41A1-37) found at Lower Fort Garry near Winnipeg. It was used inside a penitentiary in the 1880s. (a) Side view. (b) Cross section. (Drawings by D. Kappler.)

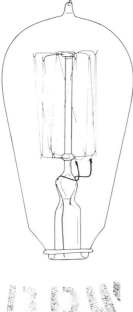

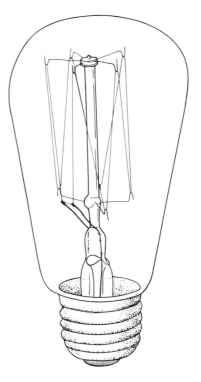

Figure 76. Incandescent lamp and customer mark (1U5A1-21). Height: 12.0 cm (without base). The single wire filament forms a cage, fastened by numerous metallic supports to a central glass tube extending the bulb. Composition: probably tungsten or tantalium. Clear bulb. Approximate date: first quarter of the 20th century. The bulb shows, engraved in acid, the initials DPW, likely those of the client, the Department of Public Works. General use impossible to determine. (Drawing by D. Kappler.)

Figure 77. Incandescent lamp (1U2A1-298). Height: 11 cm. Single wire filament forming a cage, but more open than that of the preceding lamp. It is fastened in the same way. Clear bulb. Copper screw-type base. Approximate date: first quarter of the 20th century. Use is mainly domestic, but perhaps also commercial. (Drawing by D. Kappler.)

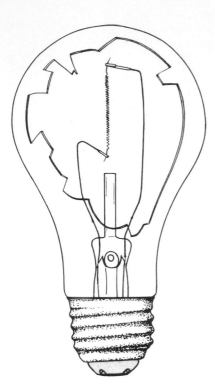

Figure 78. Modern incandescent lamp (1980). Height: 11.0 cm. The straight double-coiled filament, probably with a tungsten base, is held vertically. The bulb is glazed inside. There is an aluminum base. Power: 60 watts. Tension: 115-125 volts. Manufacturer: Westinghouse (Canada). This is the type of bulb in most widespread general use today. By comparison with the previous types, it shows how the performance of this bulb has improved with development of the filament and simplification of the construction. (Drawing by D. Kappler.)

FUNCTION

Most of the lamps can be listed under one or several of the following general functions:
1. Personal lighting (flashlight).
2. House lighting (various types of fixed and portable lights such as table lamps, ceiling lights, and so forth).
3. Urban lighting (streets, public areas, monuments).
4. Promotional lighting (illuminated signs, floodlights).

Figure 79. Modern incandescent lamp for an automobile (1U2A1-299). Height: 4.8 cm. There is a double-coiled straight tungsten-based filament and copper bayonet base. Manufacturer: General Electric (Canada). This is a model in widespread use today that has been manufactured for a long time. The inscription in ink on the cap is GECANADA/12V 21W/N17732-2/K15830/1, thus enabling us to date this model accurately. It should be noted that in Europe bayonet-type mounts are used on domestic lamps in general use. (Drawing by D. Kappler.)

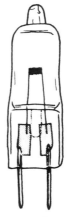

Figure 80. Modern incandescent high-powered lamp for projecting pictures (1U2A1-305). Length: 4.4 cm. There is a single large-diameter coiled filament flattened in the shape of a grid to get the largest glowing surface. The glass bulb is of a special composition that is particularly resistant to heat. The glass base forms a single unit with the bulb with two plugs for carrying the current. Power: 50 watts. Tension: 12 volts. Manufacturer: Sylvania (Federal German Republic). This lamp is one of the very many types of filament lamps designed for special use. (Drawing by D. Kappler.)

Figure 81. High-powered tubular incandescent lamp for taking photographs (1U2A1-304). Length: 11.8 cm. Filament with a large diameter for increased brightness, wound into a single, straight spiral, centred in the tube with spirals spaced approximately 1 cm apart. The filament is probably a tungsten alloy. The straight tubular bulb is of special glass that is resistant to very high temperatures. There is a central pumping opening and glass bases at each end that form a single unit with the bulb. The base contacts are set in a porcelain mantle. Tension: 120 volts. Make: Sylvania FCM (Canada). This is another example of incandescent equipment designed to give a bright light with a very small lamp. (Drawing by D. Kappler.)

Figure 82. Graphite electrode for an arc lamp (34H7C7-6). Diameter: 1.2 cm. It consists of a cylinder of a carbon-rich substance such as graphite, molded by compression in a two-part mold. One end of new electrodes was rounded or blunt. It should be noted that in the lamps that operated on direct current, two new electrodes of the same size would not be consumed at the same rate; the positive electrode would be consumed faster. The problem was solved by making the electrodes of different sizes. In addition, the end of the negative electrode would be consumed into a point whereas the other electrode would form a hollow. (Drawing by D. Kappler.)

Figure 83. Base of a modern fluorescent tube in current use (1U2A1-301). Diameter: 3.8 cm for a tube that is approximately 1 m long. The dimensions of the tubes are standardized. They can also be found ring- or U-shaped. The inside of the tubular bulb of thin glass is covered with a white powder that has fluorescent properties (the composition varies). A straight double-spiralled filament forms the hot cathode to promote the movement of electrons in the gas. A plastic base includes two contact pins. The two bases are identical. Manufacturer: Westinghouse (Canada), model F40CW. Power absorbed: 40 watts. The tube produces a cool white or neutral white light. For the past few years tubes with various spectral qualities have also been manufactured, in particular a "daylight" type that is near 4800 K which makes it possible to perceive colours accurately and another which produces a light that is richer in ultraviolet rays to speed up the growth of plants indoors. Ionized gas lamps are now in widespread use and will soon replace incandescent lamps in almost all applications. (Drawing by D. Kappler.)

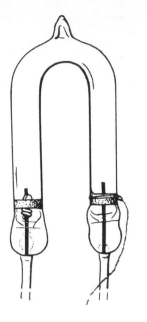

Figure 84. Ionized xenon flash bulb (1U2A1-306). Height: 5.4 cm. They are manufactured in many sizes and various shapes. The most common ones are U-or coil-shaped. The two ends are flattened to fasten the electrodes and form the base. The contact pins extend inside to form the electrodes. It should be noted that the two metallic rings at the ends of the tube are interconnected by a thin conducting band along the outside surface of the tube inside the curve. These items are part of the flash-triggering device. The flash results from an arc that jumps along the tube when a potential of several thousand volts is applied between the electrodes. The tube shown here is a flashbulb used in photography, but xenon flashbulbs are also used in signaling (lighted beacons, automatic activating of fog sirens) and scientific study of rapid movements by means of stroboscopy. (Drawing by D. Kappler.)

5. Lighting for transportation (trains, boats, automobiles, bicycles).
6. Ornamental, artistic, and recreational lighting (theatre stages, studios, coloured filters).
7. Commercial and industrial lighting (mines, hospitals, factories, stations, schools).
8. Lighting for specialized functions such as safety and signaling; black light (ultra-violet and infrared); photography (special actinic lamps, projectors, flashes); lighting in laboratory equipment.

DESCRIPTION OF EQUIPMENT AND ITS PARTS

Size

(See illustrations and the text about each type of lamp.) For the incandescent lamp it is important to note clearly the filament (shape, number, and composition if possible), the bulb, and the base. A drawing is very useful. For the arc lamp the description will be more complex, depending on the size of the remnants. An orthographic section or a blown-up view is important. For ionization lamps a schematic drawing and a list of the parts present will generally suffice.

Inscriptions

Many lamps, especially from the 20th century, have more or less indelible inscriptions. Often they give the manufacturer, the power absorbed (wattage), the tension during use (voltage), the use (for example, "hospital lamp"), the country of origin, and various characteristics ("cool white," "3200 K"). More rarely, they mention the major client ("DPW").

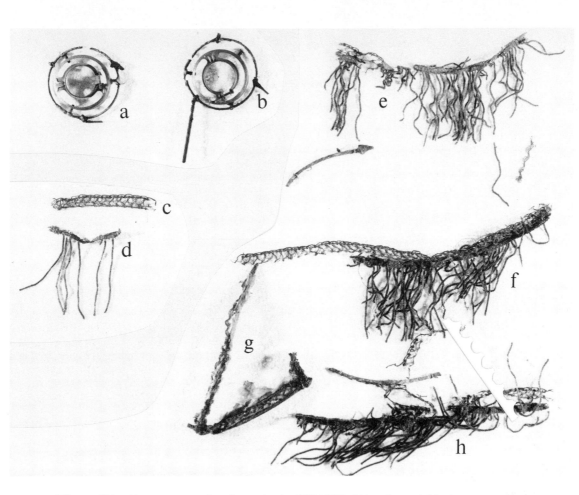

Figure 85. Fragments of a lampshade (1K40H2-34). (a and b) Bronze collars with iron hoops; (d, e, f, and h) fragments of bronze fringe; (g) pieces of silk stretched over an iron hoop. The lampshade consisted of an iron frame covered with stretched silk. Its lower edge was decorated with a fringe consisting of an ornamental band and twists of coiled bronze wire. It was suspended from the socket of the lamp by one of the collars (a) or (b) attached to the upper part of the frame. This object, which comes from Lower Fort Garry, context dating from the 1870s and 1880s, has the patent dates pressure inscribed on the two collars: PAT D. DEC. 9.90 SEPT 29.96. Based on its style, the lampshade would be suitable for a living room, a bedroom, or a private office.

BIBLIOGRAPHY

Bishop, J. Leander
1967. A History of American Manufactures from 1608 to 1860: Exhibiting .... Johnson Reprint Corporation, New York, N.Y., Vol. II. (Reprint of 3rd edition, 1868.)

Buckley, Francis
1930. "Fine Old English Glasses Part VII — Glass Candlesticks." Glass, Vol. 7 (September), pp. 356-58, 372.

Butler, Joseph T.
1967. Candleholders in America 1650-1900. A Comprehensive Collection of American and European Candle Fixtures Used in America. Crown Publishers, Inc., New York.

Carpenter, Ralph E., Jr.
1966. "Candlesticks, Sconces and Chandeliers." In Concise Encyclopedia of American Antiques, edited by Helen Comstock, Hawthorn Books Inc., New York, N.Y. pp. 564-94.

Cox, Henry Bartholomew
1979. "Hot Hairpin in a Bottle: The Beginning of Incandescence." Nineteenth Century, Vol. V, No. 3, pp. 45-49.

Cuffley, Peter
1973. A Complete Catalogue and History of Oil and Kerosene Lamps in Australia. Pioneer Design Studio Pty. Ltd., Victoria, Australia.

Curle, Alexander O.
1925-26. "Domestic Candlesticks from the Fourteenth to the end of the Eighteenth Century." Proceedings of the Society of Antiquaries of Scotland, Edinburgh, Vol. 60, pp. 183-214.

Darbee, Herbert C.
1965. "A Glossary of Old Lamps and Lighting Devices." American Association for State and Local History. Technical Leaflet 30, History News, Vol. 20, No. 8 (August), 16 pp. Nashville, Tennessee.

Davis, Pearce
1949. The Development of the American Glass Industry. Harvard University Press, Cambridge, Massachusetts.

Diamond Flint Glass Company Limited
1903-13. Lamp Chimney Catalogue and Price List of Diamond Flint Glass Co. Limited.

Dominion Glass Company Limited
Post 1913. Lamp Chimney Catalogue No. 14, Dominion Glass Company Limited.

Dubuisson, Bernard
1968. Encyclopédie pratique de la construction et du bâtiment. Librairie Aristide Quillet, Paris. Vol. II.

Duncan, Alistair
1978. Art Nouveau and Art Deco Lighting. Simon and Schuster, New York, N.Y.

Easterbrook, W.T., and Hugh G.J. Aitken
1958. Canadian Economic History. The Macmillan Company of Canada Limited, Toronto, Ontario.

Electrician, The
1902-5. Vols. 50-54. Salisbury-Court, Fleet-Street, London. (A weekly illustrated journal of electrical engineering, industry, science, and finance.

Encyclopedia Britannica
1911. University Press, Cambridge. Eleventh Edition. 20 vols.

Freeman, Larry
1955. Light on Old Lamps. Century House, Watkins Glen, N.Y.

Gentle, Rupert, and Rachael Feild
1975. English Domestic Brass 1680-1810 and the History of its Origins. E.P. Dutton & Co., Inc., New York, N.Y.

George Worthington Co.
1916. Hardware Trade Catalogue. Cleveland, Ohio.

Gloag, John
1955. A Short Dictionary of Furniture. Studio Publications Inc. in association with Thomas Y. Crowell Co., New York, N.Y.

Gowld, Mary Earle
1974. Antique Tin and Tole Ware. Charles E. Tuttle, Rutland, Vermont.

Haynes, Edward Barrington
1959. Glass Through the Ages. Penguin Books Ltd., Harmondsworth, England.

Hayward, Arthur H.
1962. Colonial Lighting. Dover Publications, New York, N.Y. Third edition. (Original edition, B.J. Brimmer Co., Boston, Massachusetts, 1923. Revised edition, Little, Brown & Co., Boston, Massachusetts, 1927.)

Hazen, Edward
1970. The Panorama of Professions and Trades; or Every Man's Book. Century House, Watkins Glen, N.Y. Reprint. (Original edition, Uriah Hunt, Philadelphia, 1836.)

Hughes, G. Bernard
1956. English Glass for the Collector 1660-1860. Lutterworth Press, London.

Hurst, J.G.
1977. "Spanish Pottery Imported into Medieval Britain." Medieval Archaeology, Vol. XXI.

Innes, Lowell
1976. Pittsburgh Glass 1797-1891: A History and Guide for Collectors. Houghton Mifflin Company, Boston.

Kauffman, Henry J.
1968. American Copper and Brass. Thomas Nelson and Sons, Toronto, Ontario.

Kirk, R.E., and D.F. Othner
1967. Encyclopedia of Chemical Technology. Wiley Interscience, John Wiley & Sons Inc., New York, N.Y. Second edition. 22 vols.

Knapp, F.
1848. Chemical Technology; or, Chemistry, Applied to the Arts and to Manufactures. Lea and Blanchard, Philadelphia, Pennsylvania. First American edition, with notes and additions by Professor Walter R. Johnson. Vol. 1. 2 vols.

Knight, Charles
1855. Knowledge is Power. John Murray, London.

Korvemaker, E. Frank
1972. Archaeological Excavations at Fort Lennox National Historic Park, 1971. Manuscript Report, No. 101, National Historic Parks and Sites, Ottawa, Ontario.

Lafferty, James R., Sr.
1969. "The Phoenix," Dedicated to the Phoenix Glass Company. James R. Lafferty Sr., n.p.

Lindsay, J. Seymour
1970. Iron and Brass Implements of the English House. Alex Tiranti, London. (Original 1927.) pp. 41-60, Figs. 214-362.

Lovell, John, ed.
1857. The Canada Directory for 1857-58; Containing Names of Professional and Business Men, and of The Principal Inhabitants, in the Cities, Towns and Villages Throughout the Province.... John Lovell, Montreal, Quebec.

Martin, Thomas
1813. The Circle of the Mechanical Arts. Printed for Richard Rees, London.

McKearin, George S., and Helen McKearin
1948. American Glass. Crown Publishers, New York, N.Y.

Merseyside County Museums
1979. Historic Glass from Collections in North West England. Merseyside County Council, Liverpool.

Myers, Denys Peter
1978. Gaslighting in America: A Guide for Historic Preservation. U.S. Department of the Interior, Heritage Conservation and Recreation Service, Office of Archeology and Historic Preservation, Technical Preservation Services Division, Washington, D.C.

Newman, Harold
1977. An Illustrated Dictionary of Glass. Thames and Hudson Ltd., London.

O'Dea, W.T.
1958. A Short History of Lighting. Ministry of Education: Science Museum, London.

Oman, Charles
1962. English Domestic Silver. Adams & Charles Black, London. Fifth edition. pp. 169-81.
1972. Pennsylvania Glassware 1870-1904. American Historical Catalog Collection, The Pyne Press, Princeton.

Perry, David H.
1969. Out of Darkness. Rochester Museum and Science Centre, Rochester, N.Y.

Perry, Evan
1974. Collecting Antique Hardware. Double-day Co. Inc., Garden City, N.Y.

Phillips, Charles John
1960. Glass - Its Industrial Applications. Reinhold Publishing Corporation, New York, N.Y.

Pyne Press Editors
1972. Lamps & Other Lighting Devices 1850-1906. American Catalogue Collection, The Pyne Press, Princeton.

Rosenberg, N.
1969. The American System of Manufactures. Edinburgh University Press, Edinburgh. (Reprint of 1854 and 1855 reports.)

Russell, Loris S.
1968. A Heritage of Light: Lamps and Lighting in the Early Canadian Home. University of Toronto Press, Toronto, Ontario.
1976. "Early Nineteenth-Century Lighting." In Building Early America, Chilton Book Company, Radnor, Pennsylvania. Chap. 11, pp. 186-201.

Scoville, Warren C.
1948. Revolution in Glassmaking. Harvard University Press, Cambridge, Massachusetts.

Seale, William
1979(?). Recreating the Historic House Interior. American Association for State and Local History, Nashville, Tennessee. Chap. 8, p. 00

Seguin, Robert-Lionel
1967. La civilisation tradionnelle de l'habitant aux 17e et 18e Siècles. Fleur de Lys Series. Fides, Montreal.

Shand, E.B.
1958. Glass Engineering Handbook. McGraw-Hill Book Company, Inc., Toronto, Ontario.

Smith, Frank R., and Ruth E. Smith
1968. Miniature Lamps. Thomas Nelson & Sons (Canada) Limited, Toronto, Ontario.

Smith, Joseph
1975. Explanation, or Key to the Various Manufactories of Sheffield. Early American Industries Association, South Burlington, Vermont. (Original 1816.)

Thuro, Catherine M.V.
1976. Oil Lamps: The Kerosene Era in North America. Wallace-Homestead Book Co., Des Moines, Iowa.

Ure, Andrew
1848. Dictionary of Arts, Manufacturing and Mines. D. Appleton, New York, N.Y. 2 vols.

Walker, David F., and James H. Bater, eds.
1974. Industrial Development in Southern Ontario. The Department of Geography, University of Waterloo, Waterloo, Ontario.

Wakefield, Hugh
1968. "Early Victorian Styles in Glassware." Studies in Glass History and Design, VIIIth International Congress on Glass, London, 1-6 July, pp. 50-54.

Watkins, C. Malcolm
1966. "Lighting Devices." In Concise Encyclopedia of American Antiques, edited by Helen Comstock, Hawthorn Books Inc., New York, N.Y.

Wells, Stanley
1975. Period Lighting. Pelham Books, London.

Wood, Vallance Limited
1911. Wholesale Shelf and Heavy Hardware Bar Iron and Steel. (Trade catalogue.) Winnipeg, Manitoba.